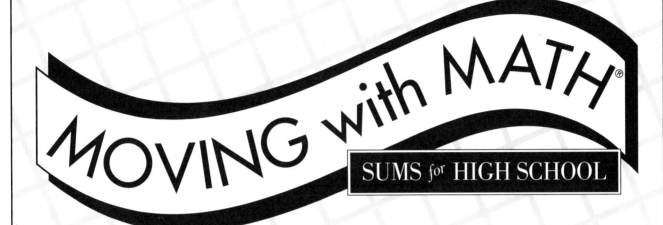

MOVING with MATH®

SUMS for HIGH SCHOOL

Success Using Math Standards for High School:
Preparation for Success

Part I

Math Teachers Press, Inc.

A Complete
Program for
Math Success in
High School

Number Sense and Operations

Table of Contents

A Place Value Chart to Billions

The national debt of the United States in 1991 is shown in the place value chart.

		Billions			Millions			Thousands			Units		
Hundred Billions	Ten Billions	Billions	Hundred Millions	Ten Millions	Millions	Hundred Thousands	Ten Thousands	Thousands	Hundreds	Tens	Ones		
		3	6	8	3	0	0	0	0	0	0		

Look at the number in the place value chart above.
What digit is in each place?

1. billions _____   2. hundred millions _____   3. ten millions _____

Write the place name and value of each underlined digit.

	Place Name	Value
4. 6,473,758,216	*one billions*	*6,000,000,000*
5. 26,000,000,000		
6. 761,000,000,000		
7. 5,000,000		
8. 5,000		

For the number 793,462,105,800 write the digit that is in each place.

9. tens _____   10. thousands _____

11. hundred thousands _____   12. millions _____

13. billions _____   14. ten millions _____

15. hundred billions _____   16. hundred millions _____

17. hundreds _____   18. ten billions _____

A Place Value Chart to Billions

This place value chart names the first twelve places.
Notice how the pattern of "one - ten - hundred" repeats in groups of three.

HUNDRED BILLIONS	TEN BILLIONS	BILLIONS	HUNDRED MILLIONS	TEN MILLIONS	MILLIONS	HUNDRED THOUSANDS	TEN THOUSANDS	THOUSANDS	HUNDREDS	TENS	ONES
BILLIONS			MILLIONS			THOUSANDS			UNITS		
		3	4	6	2	0	9	7	5	1	8

The number shown is 3 billions or 3,000,000,000

4 hundred millions 400,000,000

6 ten millions 60,000,000

2 one millions 2,000,000

0 hundred thousands 000,000

9 ten thousands 90,000

7 one thousands 7,000

5 hundreds 500

1 ten 10

8 ones 8

The sum of the values is the number....................................3,462,097,518

Look at the number in the place value chart above. What digit is in each place?

1. ten thousands _____   **2.** hundred millions _____   **3.** one billions _____

Write a numeral to show the value of:

4. 2 millions, 9 ten thousands, 4 hundreds   _____

5. 4 hundred millions, 2 thousands, 7 tens   _____

Write each number in expanded form.

6. 8,005,070   = _____ + _____ + _____

7. 370,400,000   = _____ + _____ + _____

8. 64,050,700   = _____ + _____ + _____ + _____

9. 4,003,700,008   = _____ + _____ + _____ + _____

10. 2,000,346   = _____ + _____ + _____ + _____

Reading and Evaluating Numbers in Exponential Form

The sides of a square are 3 units. Write the area in exponential form and find the area.

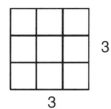

3

3

Area = 3 × 3 or 3²
= 9 square units

3² is read "3 to the second power" or "3 squared".

The sides of a cube are 2 units. Write the volume in exponential form and find the volume.

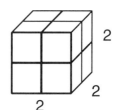

2

2

2

Volume = 2 × 2 × 2 or 2³
= 8 cubic units

2³ is read "2 to the third power" or "2 cubed".

Write the numeral in exponential form.

1. 2 squared

2. 6 cubed

3. 10 to the second power

_____

_____

_____

4. 2 to the fifth power

5. 7 squared

6. 10 cubed

_____

_____

_____

Write the words for each expression.

7. 6^3 _____

8. 8^2 _____

9. 2^4 _____

10. 2^5 _____

11. 10^3 _____

12. e^3 _____

Write in factored form and evaluate.

13. 7^3 = __×__×__ = _____

14. 10^2 = ___×_____ = ___

15. 8^2 = ___×_____ = ___

16. 6^2 = _____ = _____

17. 2^4 = _____ = ___

18. 12^2 = _____ = ___

19. 2^2 = _____ = _____

20. 6^3 = _____ = ___

21. 2^3 = _____ = ___

22. 10^3 = _____ = _____

23. 4^2 = _____ = ___

24. 10^4 = _____ = ___

25. 5^3 = _____ = _____

26. 2^5 = _____ = ___

27. 3^3 = _____ = ___

Write each number in exponential form with a base of 10.

28. 100 = _____

29. 1000 = _____

30. 10 = _____

31. 10,000 = _____

32. 100,000 = _____

33. 1,000,000 = _____

34. What is the pattern relating to the number in problems 28–33?

What is the largest number that can be written with three digits?

The Square Root of a Number

<table>
<tr><td>

The **square** of a number is the total number of unit squares needed to form a larger square with sides equal to the given number.

6 squared is 36

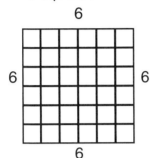

</td><td>

The **square root** of a number is the opposite of the square of a number. The square root is the length of the missing side when you know the total number of unit squares in the larger square.

What is the square root of 16?

?

| 16 | ? |

There will be 4 units on each side. The square root of 16 is 4.

We write: $\sqrt{16} = 4$

</td></tr>
</table>

The area of the square is given.
How many units on the side of each square?

		Number	Square root
		1	1
		4	2
		9	3
		16	4
		25	5

1.

256

s = _____

2.

529

s = _____

Find the square of each number.

3. $6^2 =$ _____

4. $16^2 =$ _____

5. $20^2 =$ _____

6. $13^2 =$ _____

Find the square root.

7. $\sqrt{36} =$ _____

8. $\sqrt{64} =$ _____

9. $\sqrt{121} =$ _____

10. $\sqrt{324} =$ _____

11. $\sqrt{625} =$ _____

12. $\sqrt{400} =$ _____

13. $\sqrt{3^2 + 4^2} =$ _____

14. $\sqrt{6^2 + 8^2} =$ _____

Number	Square root
1	1
4	2
9	3
16	4
25	5
36	6
49	7
64	8
81	9
100	10
121	11
144	12
169	13
196	14
225	15
256	16
289	17
324	18
361	19
400	20
441	21
484	22
529	23
576	24
625	25

Estimating Square Roots

You have 29 carpet squares. You want to put them together to make the largest square rug possible without cutting any of the squares.

How many squares will there be on each side?

Can a square be built with 29 square units?
The longest perfect square that can be built
with 29 square units is a square of 5.

Sometimes the square root
is not a whole number.

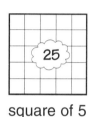

25

The square root of 29 comes
between 5 and 6.

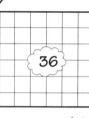

36

square of 5

square of 6

Find the $\sqrt{}$ of the numbers below. If the number is not a perfect square, tell which whole numbers the $\sqrt{}$ is between. Use your unit cubes to help.

1. $\sqrt{12}$

2. $\sqrt{31}$

3. $\sqrt{25}$

4. $\sqrt{17}$

5. $\sqrt{49}$

6. $\sqrt{153}$

Plot and label each point on the number line below. Show between which two numbers the following square roots can be found.

7. $\sqrt{71}$

8. $\sqrt{21}$

9. $\sqrt{13}$

10. $\sqrt{52}$

11. $\sqrt{94}$

12. $\sqrt{11}$

Place Value with Exponents

Our system of numerals is a place value system.
This place value chart shows the names of the first six place values.
Each place value is written in numbers and with exponents.
Find the pattern of the exponents and complete the chart.

Hundred Thousands	Ten Thousands	One Thousands	Hundreds	Tens	Ones
_____	_____	1000	100	_____	1
10x10x10x10x10	10x10x10x10	10x10x10	10x10	10x1	
$10^{\square}$	$10^{\square}$	10^3	10^2	$10^{\square}$	$10^{\square}$

1. Ten to the first power or 10^1 = _____

2. Ten to the zero power or 10^0 = _____

3. $400 = \underline{4 \times 100} = \underline{4 \times 10^2}$ 4. $7,000 = \underline{7 \times 1000} = \underline{7 \times 10^3}$

5. $60 =$ _____ = _____   6. $80,000 =$ _____ = _____

7. $900 =$ _____ = _____   8. $500,000 =$ _____ = _____

9. $2 =$ _____ = _____   10. $3,000,000 =$ _____ = _____

Write a standard number for each of the following:

11. $3 \times 10^4 = 3 \times 10,000 = 30,000$ 12. $7 \times 10^2 =$

13. $6 \times 10^5 =$ 14. $5 \times 10^3 =$

15. $8 \times 10^1 =$ 16. $2 \times 10^0 =$

17. Each place value is _____ times as great as the next place value to the right.

18. Tell about the relationship between the number of zeros in a place value name such as 1000 or 10,000 to the exponent.

Scientific Notation

Scientists often work with very large numbers. They use an abbreviated method of writing numbers known as **scientific notation.** To express any number in scientific notation, write it as the product of a decimal from 1 to 9 multiplied by a power of 10.

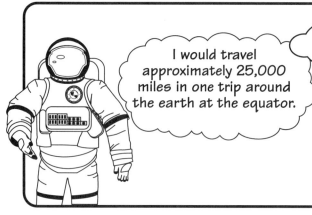

I would travel approximately 25,000 miles in one trip around the earth at the equator.

Think about the place value chart.
25,000 = 2.5 ten thousands

2 •	5	0	0	0
10^4	10^3	10^2	10^1	10^0

$$25,000 = 2.5 \times 10^4$$

Write each number as a power of ten.

1. $10,000 = 10^4$

2. $1,000,000 = $ _____

3. $100,000 = $ _____

4. $10 = $ _____

5. $100 = $ _____

6. $1 = $ _____

7. $1000 = $ _____

8. $10,000,000 = $ _____

9. $100,000,000 = $ _____

Write each number in scientific notation. The first one has been done for you.

10. $3500 = \underline{3.5} \cdot \underline{1,000}$

  $= \underline{3.5} \cdot \underline{10^3}$

11. $460,000 = $ _____ $\times$ _____

  $= $ _____ $\times$ _____

12. $14,000,000 = $ _____ $\cdot$ _____

  _____ $\cdot$ _____

13. $89,300 = $ _____ $\times$ _____

  _____ $\times$ _____

14. $50,600 = $ _____ $\cdot$ _____

  _____ $\cdot$ _____

15. $975.2 = $ _____ $\times$ _____

  $= $ _____ $\times$ _____

16. The sun is approximately 4,500,000,000 years old. Write the age of the sun in scientific notation.

  _____

17. The earth is approximately 150,000,000 km from the sun. Write the distance in scientific notation.

  _____

18. The diameter of a giant star is about 1,700,000 miles. Express this number in scientific notation.

  _____

Scientific Notation with Negative Exponents

Scientists write very small numbers in a quick form called **scientific notation.** A numeral is changed to this quick form by writing it as the product of a number from 1 to 9 times a power of 10. This expanded place value chart shows the standard and exponential form for whole numbers and decimal fractions. Notice that the exponent of numbers less than 1 are negative.

place name	Thousands	Hundreds	Tens	Ones	Tenths	Hundredths	Thousandths	Ten thousandths
standard notation	1,000	100	10	1	$\frac{1}{10}$	$\frac{1}{100}$	$\frac{1}{1000}$	$\frac{1}{10000}$
exponential form	10^3	10^2	10^1	10^0	10^{-1}	10^{-2}	10^{-3}	10^{-4}

$$0.00635 \quad = 6.35 \times \frac{1}{1000}$$
$$= 6.35 \times 10^{-3}$$

The decimal point has been moved 3 places to the right.

Write in scientific notation.

1. $0.0146 = 1.46 \times \frac{1}{100}$
$$= 1.46 \times 10^{-2}$$

2. $0.234 = $ _____
$$= $$ _____

3. $0.00526 = $ _____
$$= $$ _____

4. $0.407 = $ _____
$$= $$ _____

5. $0.0079 = $ _____
$$= $$ _____

6. $0.000845 = $ _____
$$= $$ _____

7. $0.00706 = $ _____
$$= $$ _____

8. $0.00091 = $ _____
$$= $$ _____

9. $0.00793 = $ _____
$$= $$ _____

Write in standard notation.

10. $3.14 \times 10^{-4} = 3.14 \times \frac{1}{10000}$
$$= \underline{3.14 \div 10,000}$$
$$= \underline{.000314}$$

11. $2.4 \times 10^{-1} = $ _____
$$= $$ _____
$$= $$ _____

12. $1.7 \times 10^{-2} = $ _____
$$= $$ _____
$$= $$ _____

13. $4.6 \times 10^{-3} = $ _____
$$= $$ _____
$$= $$ _____

14. $7.3 \times 10^{-2} = $ _____
$$= $$ _____
$$= $$ _____

15. $6.1 \times 10^{-1} = $ _____
$$= $$ _____
$$= $$ _____

16. $2.03 \times 10^{-4} = $ _____
$$= $$ _____
$$= $$ _____

17. $3.1416 \times 10^{-4} = $ _____
$$= $$ _____
$$= $$ _____

Multiplying Exponents with a Common Base

If:
$7^2 = 7 \times 7$, and
$7^4 = 7 \times 7 \times 7 \times 7$, then
$7^2 \times 7^4 = 7 \times 7 \times 7 \times 7 \times 7 \times 7$.
We can write this as 7^6 because
we multiply 7 six times.
Thus, $7^2 \times 7^4 = 7^6$

$2^3 = \underline{} \times \underline{} \times \underline{}$

$2^4 = \underline{} \times \underline{} \times \underline{} \times \underline{}$

$2^3 \times 2^4 = \underline{} \times \underline{} \times \underline{} \times \underline{} \times \underline{} \times \underline{} \times \underline{}$

How can $2^3 \times 2^4$ be written using one base and one exponent? _____

What is the pattern for multiplying exponentials with a common base?

_____

Simplify the following expressions. Do not evaluate.

1. $7^3 \times 7^1 =$ _____

2. $3^4 \times 3^6 =$ _____

3. $4^2 \times 4^7 =$ _____

4. $5^6 \times 5^0 =$ _____

5. $11^3 \times 11^9 =$ _____

6. $15^0 \times 15^7 =$ _____

7. $9^1 \times 9^3 \times 9^5 =$ _____

8. $6^0 \times 6^2 \times 6^4 =$ _____

9. $4^2 \times 3^2 \times 4^3 =$ _____

10. $7^2 \times 5^4 \times 5^6 =$ _____

11. $2^5 \times 2^1 \times 9^0 =$ _____

12. $4^3 \times 7^2 \times 4^4 \times 7^5 =$ _____

Dividing Exponents with a Common Base

What is $3^5 \div 3^3$?

$3^5 = 3 \times 3 \times 3 \times 3 \times 3$

$3^3 = 3 \times 3 \times 3$

$\dfrac{3^5}{3^3} = \dfrac{3 \times 3 \times 3 \times 3 \times 3}{3 \times 3 \times 3}$

Since $\dfrac{3}{3} = 1$, you can simplify by first canceling.

$$\dfrac{\cancel{3} \times \cancel{3} \times \cancel{3} \times 3 \times 3}{\cancel{3} \times \cancel{3} \times \cancel{3}} = 3^2 = 9$$

Similarly,

$4^4 = 4 \times 4 \times 4 \times 4$

$4^3 = 4 \times 4 \times 4$

$\dfrac{4^4}{4^3} = \dfrac{\cancel{4} \times \cancel{4} \times \cancel{4} \times 4}{\cancel{4} \times \cancel{4} \times \cancel{4}} = 4^1$

What is the pattern for dividing exponents with the same base?

_____

Simplify the following expressions. Leave in exponential form.

1. $\dfrac{7^6}{7^3} = \dfrac{\cancel{7} \times \cancel{7} \times \cancel{7} \times 7 \times 7 \times 7}{\cancel{7} \times \cancel{7} \times \cancel{7}} = 7^3$ 2. $\dfrac{5^9}{5^4} = $ _____   3. $\dfrac{3^7}{3^2} = $ _____

4. $2^6 \div 2^5 = $ _____   5. $8^3 \div 8^2 = $ _____   6. $12^6 \div 12^4 = $ _____

7. $\dfrac{3^3 \times 3^4}{3^5} = $ _____   8. $\dfrac{3^2 \times 4^2}{4^2} = $ _____   9. $\dfrac{7^1 \times 7^6}{7^3} = $ _____

10. $\dfrac{3^2 \times 4^3}{3^2 \times 4^2} = $ _____   11. $\dfrac{8^4 \times 5^2}{5^2 \times 8^2} = $ _____   12. $\dfrac{9^4 \times 2^2}{9^2 \times 2^3} = $ _____

Simplifying Numbers in Exponential Form

You know that:

$$\frac{4^5}{4^3} = \frac{\cancel{4} \times \cancel{4} \times \cancel{4} \times 4 \times 4}{\cancel{4} \times \cancel{4} \times \cancel{4}} = 4 \times 4 = 4^2$$

Pattern:

$$4^{5-3} = 4^2$$

What happens to?

$$\frac{4^3}{4^5} = \frac{\cancel{4} \times \cancel{4} \times \cancel{4}}{\cancel{4} \times \cancel{4} \times \cancel{4} \times 4 \times 4} = \frac{1}{4 \times 4} \text{ or } \frac{1}{4^2}$$

$$4^{3-5} = 4^{-2}$$

Pattern: To divide numbers in exponential form, _____ the exponents.

Write the following with positive exponents.

1. $4^{-6} = \dfrac{1}{4^6}$

2. $10^{-2} =$ _____

3. $9^{-4} =$ _____

4. $13^{-5} \times 7^2 =$ _____

5. $6 \times 2^{-3} =$ _____

6. $4^{-1} \times 5^{-3} =$ _____

Evaluate the following expressions. Make sure your answers have positive exponents. *Show your work.*

7. $\dfrac{2^2}{2^3} =$

8. $\dfrac{8^4}{8^9} =$

9. $\dfrac{10^3}{10^5} =$

10. $\dfrac{3^4}{3^7} =$

11. $\dfrac{6^3 \times 6^5}{6^{11}} =$

12. $\dfrac{7^4 \times 7^3}{7^8} =$

Problem Solving: Find the Pattern

These fraction bars have been sorted into groups by some way they are alike or similar. What is the similarity?

1.

Similarity: <u>divided into 2 parts</u>

2.

Similarity: _____

_____

3.

Similarity: _____

_____

4.

Similarity: _____

_____

5.

Similarity: _____

_____

6.

Similarity: _____

_____

7.

Similarity: _____

_____

8.

Similarity: _____

_____

Problem Solving: Make a Table

Find all the sets of equivalent fractions from a set of fraction bars.

Lowest Terms	Equivalent Fractions in Higher Terms					
	$\frac{1}{3}$'s	$\frac{1}{4}$'s	$\frac{1}{5}$'s	$\frac{1}{6}$'s	$\frac{1}{10}$'s	$\frac{1}{12}$'s
1. $\frac{1}{2}$		$\frac{2}{4}$		$\frac{3}{6}$	$\frac{5}{10}$	$\frac{6}{12}$
2. $\frac{2}{2}$	$\frac{3}{3}$	$\frac{4}{4}$	$\frac{5}{5}$	$\frac{6}{6}$	$\frac{10}{10}$	$\frac{12}{12}$
3. $\frac{1}{3}$						
4. $\frac{2}{3}$						
5. $\frac{1}{4}$						
6. $\frac{3}{4}$						
7. $\frac{1}{5}$						
8. $\frac{2}{5}$						
9. $\frac{3}{5}$						
10. $\frac{4}{5}$						

Lowest Terms	Equivalent Fractions in Higher Terms					
	$\frac{1}{3}$'s	$\frac{1}{4}$'s	$\frac{1}{5}$'s	$\frac{1}{6}$'s	$\frac{1}{10}$'s	$\frac{1}{12}$'s
11. $\frac{1}{6}$						
12. $\frac{5}{6}$						
13. $\frac{1}{10}$						
14. $\frac{3}{10}$						
15. $\frac{7}{10}$						
16. $\frac{9}{10}$						
17. $\frac{1}{12}$						
18. $\frac{5}{12}$						
19. $\frac{7}{12}$						
20. $\frac{11}{12}$						

21. A fraction is in lowest terms when _____.

Adding Unlike Fractions

Lisa did $\frac{1}{4}$ of her weekly practice on Monday. She did $\frac{1}{6}$ of her practice on Tuesday. How much of her weekly practice time has she completed?

The least common multiple for 4 and 6 is 12.

$\frac{1}{4} + \frac{1}{6} = ?$

$$\frac{1}{4} = \frac{3}{12}$$

$$+ \frac{1}{6} = \frac{2}{12}$$

$$\frac{5}{12}$$

Lisa has completed $\frac{5}{12}$ of her practice time.

Add.

1. $\frac{1}{3}$

$+ \frac{1}{2}$

2. $\frac{1}{5}$

$+ \frac{1}{2}$

3. $\frac{2}{3}$

$+ \frac{1}{4}$

4. $\frac{3}{10}$

$+ \frac{2}{5}$

5. $\frac{1}{5}$

$+ \frac{2}{3}$

6. $\frac{1}{5}$

$+ \frac{3}{4}$

7. $\frac{1}{8}$

$+ \frac{1}{2}$

8. $\frac{1}{9}$

$+ \frac{2}{3}$

9. $\frac{1}{3}$

$+ \frac{1}{12}$

10. $\frac{3}{10}$

$+ \frac{4}{10}$

11. $\frac{1}{6}$

$+ \frac{3}{4}$

12. $\frac{5}{14}$

$+ \frac{2}{7}$

13. Jane ate $\frac{5}{12}$ of a candy bar and Ray ate $\frac{3}{8}$ of the same candy bar. Explain if this is possible.

14. Jess ate $\frac{1}{4}$ of a pizza. Jack ate $\frac{1}{3}$ of the same pizza. Dennis ate $\frac{1}{2}$ of the same pizza. Explain if this is possible.

Subtracting Unlike Fractions

Mr. Gruidl began a trip with a gas tank $\frac{3}{4}$ full. When he was about $\frac{1}{2}$ of the way, his gas tank showed $\frac{1}{8}$ full. How much of his tank had he used?

Will he have enough gas to finish the trip?

The LCM for 4 and 8 is 8.

$$\frac{3}{4} = \frac{6}{8}$$
$$-\frac{1}{8} = -\frac{1}{8}$$
$$\frac{5}{8}$$

He used $\frac{5}{8}$ of his tank to go $\frac{1}{2}$ of the trip.

He will **not** have enough gas.

Subtract. Simplify your answer.

1. $\frac{1}{2}$
  $-\frac{1}{4}$

2. $\frac{2}{3}$
  $-\frac{1}{2}$

3. $\frac{1}{2}$
  $-\frac{2}{5}$

4. $\frac{1}{2}$
  $-\frac{1}{3}$

5. $\frac{5}{6}$
  $-\frac{3}{4}$

6. $\frac{7}{8}$
  $-\frac{1}{4}$

7. $\frac{8}{10}$
  $-\frac{1}{2}$

8. $\frac{7}{10}$
  $-\frac{3}{5}$

9. $\frac{5}{12}$
  $-\frac{1}{3}$

10. $\frac{7}{12}$
  $-\frac{1}{2}$

11. $\frac{9}{10}$
  $-\frac{3}{5}$

12. $\frac{2}{3}$
  $-\frac{5}{9}$

13. A recipe for cookies calls for $\frac{2}{3}$ cup of brown sugar and $\frac{1}{2}$ cup of white sugar. How much of both sugars is needed?

14. Carol jogged $\frac{5}{8}$ of a mile in the morning and $\frac{3}{4}$ of a mile in the afternoon. How much farther did she jog in the afternoon?

Use the pictures to find the products.

1.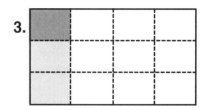

$\dfrac{1}{2}$ of $\dfrac{1}{3}$ =

2.

$\dfrac{1}{3}$ of $\dfrac{1}{2}$ =

3.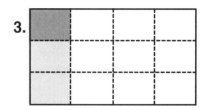

_____ of _____ = _____

Multiply. Write the product in lowest terms.

4. $\dfrac{1}{6} \times \dfrac{1}{3}$ =

5. $\dfrac{1}{4} \times \dfrac{1}{4}$ =

6. $\dfrac{1}{5} \times \dfrac{1}{2}$ =

7. $\dfrac{2}{3} \times \dfrac{2}{3}$ =

8. $\dfrac{3}{10} \times \dfrac{3}{7}$ =

9. $\dfrac{4}{5} \times \dfrac{3}{5}$ =

10. $\dfrac{1}{6} \times \dfrac{3}{4}$ =

11. $\dfrac{7}{9} \times \dfrac{1}{2}$ =

12. $\dfrac{5}{12} \times \dfrac{1}{3}$ =

13. $\dfrac{1}{2} \times \dfrac{3}{16}$ =

14. $\dfrac{2}{3} \times \dfrac{2}{5}$ =

15. $\dfrac{3}{4} \times \dfrac{1}{3}$ =

16. Gene made a cake and cut it into four equal parts. He then cut each part into 3 equal pieces and ate one of the pieces. What fraction of the cake did he eat?

_____

17. Jesse had $\dfrac{5}{8}$ of a pizza. He ate $\dfrac{1}{3}$ of his pizza after school. What fraction of the whole pizza was eaten after school?

_____

18. A recipe calls for $\dfrac{3}{4}$ cup sugar. If only half a recipe is made, how much sugar should be used?

_____

19. A recipe calls for $\dfrac{2}{3}$ cup of powdered sugar. If only half a recipe is made, how much powdered sugar should be used?

_____

Multiplying a Fraction by a Whole Number

Amanda used $\frac{1}{2}$ of a dozen eggs to make a sponge cake. How many eggs did she use?

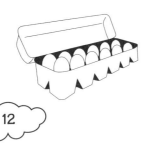

1 dozen means 12

$\frac{1}{2}$ of 12 = ?

Here are four different ways to find the answer. Can you explain each method?

1.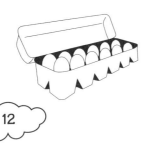

2. $2\overline{)12}$ with 6 above

3. $\frac{1}{2} \times \frac{12}{1} = \frac{1 \times 12}{2 \times 1} = \frac{12}{2} = 6$

4. $\frac{1}{2} = \frac{2}{4} = \frac{3}{6} = \frac{4}{8} = \frac{5}{10} = \frac{6}{12} = 6$ eggs

Multiply. Simplify your answer.

1. $\frac{1}{3}$ of 12 =

2. $\frac{1}{4}$ of 12 =

3. $\frac{1}{6}$ of 12 =

4. $\frac{1}{5}$ of 10 =

5. $\frac{1}{3}$ of 30 =

6. $\frac{1}{4}$ of 16 =

7. $\frac{1}{3}$ of 18 =

8. $\frac{1}{5}$ of 100 =

9. $\frac{1}{4}$ of 40 =

10. $\frac{1}{8}$ of 14 =

11. $\frac{1}{6}$ of 48 =

12. $\frac{1}{4}$ of 100 =

13. In a class of 24 students, $\frac{1}{3}$ of them were boys. How many were boys?

14. Sam bought 8 pounds of hamburger. $\frac{1}{4}$ of it was fat and water. How much was fat and water?

15. Angela had 30¢. She spent $\frac{1}{3}$ of it on a can of juice. How much did she have left?

16. The gas tank holds 20 gallons. The tank is $\frac{1}{4}$ full. About how many gallons are in the tank?

Renaming Improper Fractions and Mixed Numbers

> To change improper fractions and mixed numbers, use these patterns.

Pattern for renaming an improper fraction as a mixed number: divide the numerator by the denominator. $$\frac{5}{3} = 3\overline{)5}\ \ \ \begin{array}{r}1\frac{2}{3}\\[-2pt]\hline-3\\\hline2\end{array}$$	To rename a mixed number as an improper fraction: multiply the denominator times the whole number. Then add the result to the original numerator. The denominator stays the same. $$1\frac{2}{3} = \frac{(3 \times 1) + 2}{3} = \frac{5}{3}$$

Change each improper fraction to a mixed number.

1. $\dfrac{3}{2}$ _____

2. $\dfrac{5}{4}$ _____

3. $\dfrac{4}{3}$ _____

4. $\dfrac{7}{5}$ _____

5. $\dfrac{11}{6}$ _____

6. $\dfrac{15}{8}$ _____

7. $\dfrac{9}{7}$ _____

8. $\dfrac{13}{9}$ _____

Change each whole number to an improper fraction.

9. $1 = \dfrac{}{2}$

10. $1 = \dfrac{}{6}$

11. $1 = \dfrac{}{3}$

12. $1 = \dfrac{}{10}$

13. $1 = \dfrac{}{8}$

14. $1 = \dfrac{}{5}$

15. $1 = \dfrac{}{9}$

16. $1 = \dfrac{}{12}$

Change each mixed number to an improper fraction.

17. $1\dfrac{1}{4} =$ _____

18. $1\dfrac{5}{8} =$ _____

19. $1\dfrac{2}{3} =$ _____

20. $1\dfrac{5}{6} =$ _____

21. $1\dfrac{3}{5} =$ _____

22. $1\dfrac{2}{7} =$ _____

23. $1\dfrac{4}{9} =$ _____

24. $1\dfrac{3}{10} =$ _____

Multiplying with Mixed Numbers

You can multiply a mixed number times any other kind of fraction if you first change the mixed number to an improper fraction.

mixed number × whole number	mixed number × proper fraction	mixed number × mixed number
$2\frac{1}{4} \times 8$	$3\frac{1}{2} \times \frac{3}{4}$	$2\frac{2}{3} \times 3\frac{1}{4}$
$\frac{9}{4} \times \frac{8}{1}$ (cancel)	$\frac{7}{2} \times \frac{3}{4}$ (simplify)	$\frac{8}{3} \times \frac{13}{4}$ (simplify)
$\frac{9}{\,_1 4} \times \frac{\overset{2}{8}}{1} = 18$	$\frac{21}{8} = 2\frac{5}{8}$	$\frac{\overset{2}{8}}{3} \times \frac{13}{\underset{1}{4}} = \frac{26}{3} = 8\frac{2}{3}$ (cancel)

Multiply. Simplify.

1. $2\frac{1}{2} \times 2\frac{1}{2} =$ **2.** $3\frac{3}{4} \times 1\frac{2}{3} =$ **3.** $3 \times 4\frac{2}{5} =$ **4.** $4 \times 2\frac{5}{8} =$

5. $1\frac{1}{3} \times 3\frac{1}{2} =$ **6.** $1\frac{5}{8} \times 3\frac{1}{3} =$ **7.** $10 \times 4\frac{2}{5} =$ **8.** $1\frac{2}{5} \times 1\frac{7}{8} =$

9. $2\frac{2}{3} \times 5\frac{1}{2} =$ **10.** $3\frac{1}{3} \times 5\frac{1}{9} =$ **11.** $3\frac{3}{4} \times \frac{2}{5} =$ **12.** $15 \times 3\frac{2}{3} =$

13. A recipe for scrambled eggs calls for $1\frac{1}{2}$ cups of milk. How much milk would be used to make half a recipe?

_____

14. A meat loaf recipe calls for $\frac{3}{4}$ cup bread crumbs. How much bread crumbs should be used for one and one-half recipes?

_____

15. A bird house requires $5\frac{1}{3}$ feet of lumber. How many feet will be needed for two bird houses?

_____

16. A roll of crepe paper is $1\frac{2}{3}$ yards long. If $3\frac{1}{2}$ rolls were used to decorate a school float, how many yards of crepe paper were used?

_____

Problem Solving: Find the Pattern to Divide Fractions

Elaine's mother owns a coffee shop featuring specialty pies. She cuts her pies into sixths to serve individual pieces. She has $\frac{1}{2}$ of an apple pie. How many pieces of pie will she have?

We need to find how many $\frac{1}{6}$'s there are in $\frac{1}{2}$.

 ← Use a fraction bar to show $\frac{1}{2}$ of a pie.

 ← Now put a $\frac{1}{6}$ fraction bar under it.

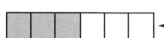 ← Think: There would be three parts shaded if the pie was cut into sixths.

apple pie

$\frac{1}{2} \div \frac{1}{6} = 3$

Now find how many sixths would be in a $\frac{1}{3}$ of a pie.

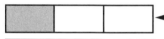 ← Use a fraction bar to show $\frac{1}{3}$ of a pie.

 ← Now put a $\frac{1}{6}$ fraction bar under it.

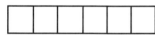 ← Shade in the number of pieces.

$\frac{1}{3} \div \frac{1}{6} = \underline{}$

Look at the fraction bars and the written problems. What is the pattern for dividing fractions? _____

Use your fraction bars to find how many sets of the smaller fraction are in the larger fraction.

1. How many $\frac{1}{12}$'s are in $\frac{1}{4}$?

   $\frac{1}{4} \div \frac{1}{12} =$ _____

2. How many $\frac{1}{12}$'s are in $\frac{1}{3}$?

   $\frac{1}{3} \div \frac{1}{12} =$ _____

3. How many $\frac{1}{12}$'s are in $\frac{5}{6}$?

   $\frac{5}{6} \div \frac{1}{12} =$ _____

4. How many $\frac{1}{10}$'s are in $\frac{3}{5}$?

   $\frac{3}{5} \div \frac{1}{10} =$ _____

5. How many $\frac{1}{10}$'s are in 1?

   $1 \div \frac{1}{10} =$ _____

6. How many $\frac{1}{5}$'s are in $\frac{1}{2}$?

   $\frac{1}{2} \div \frac{1}{5} =$ _____

Dividing by a Fraction: Multiply by the Reciprocal

You can use fraction bars to see that the given rule works.

$$2 \div \frac{1}{4} =$$

$$\frac{1}{4} \div \frac{1}{8} =$$

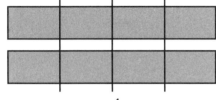

$$2 \times \frac{4}{1} = 8$$

$$\frac{1}{4} \times \frac{8}{1}$$

Note the cancelling.

$$\frac{1}{\cancel{1}\cancel{4}} \times \frac{\cancel{8}^2}{1} = \frac{2}{1} = 2$$

Rule: To divide by a fraction, multiply by its reciprocal. The reciprocal of number is the number inverted (turned upside down).

Write the reciprocal of each number.

1. $\dfrac{1}{3}$ _____

2. $\dfrac{1}{10}$ _____

3. $\dfrac{2}{3}$ _____

4. $\dfrac{3}{4}$ _____

5. $\dfrac{1}{8}$ _____

6. $\dfrac{1}{6}$ _____

7. $\dfrac{4}{5}$ _____

8. $\dfrac{3}{7}$ _____

Divide using the rule. Simplify all answers.

9. $6 \div \dfrac{1}{3} =$

10. $7 \div \dfrac{1}{8} =$

11. $8 \div \dfrac{1}{5} =$

12. $4 \div \dfrac{1}{5} =$

13. $\dfrac{1}{4} \div \dfrac{1}{3} =$

14. $\dfrac{1}{2} \div \dfrac{2}{3} =$

15. $\dfrac{3}{4} \div \dfrac{4}{5} =$

16. $\dfrac{1}{2} \div \dfrac{1}{5} =$

17. $\dfrac{9}{10} \div \dfrac{1}{3} =$

18. $\dfrac{2}{3} \div \dfrac{4}{5} =$

19. $\dfrac{7}{8} \div \dfrac{7}{16} =$

20. $\dfrac{2}{5} \div \dfrac{3}{10} =$

21. How many $\frac{2}{3}$ cup servings are there in a container holding 8 cups of juice?

_____

22. How many $\frac{3}{4}$ ounce servings are there in 12 ounces of nuts?

_____

Decimal Fraction Place Value: Tenths and Hundredths

A decimal number can be shown with a model made of blocks.

The flat square is a model of the whole number 1.

The rod is $\frac{1}{10}$ or 0.1 or "one-tenth" of the whole number 1.

The unit cube is $\frac{1}{100}$ or 0.01 or "one-hundredth" of the whole number 1.

The value of each block names its place on a place value mat.

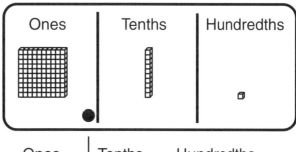

Ones	Tenths	Hundredths
2	4	8

Write each number as a decimal fraction.

1. =

2.   =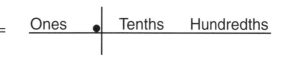

3. 5 ones, 6 tenths, 2 hundredths = _____ • _____ _____

4. 7 ones, 4 tenths, 3 hundredths = _____ • _____ _____

5. 2 ones, 7 tenths, 3 hundredths = _____ • _____ _____

6. 9 ones, 4 tenths, 2 hundredths = _____ • _____ _____

7. 2 ones, 13 tenths, 4 hundredths = _____ • _____ _____

8. 6 ones, 4 tenths, 12 hundredths = _____ • _____ _____

9. 3 ones, 0 tenths, 5 hundredths = _____ • _____ _____

10. 8 ones, 7 hundredths = _____ • _____ _____

Adding and Subtracting Decimals: Hundredths

Your mother gives you the money she saves by using coupons when she shops. During one week, she saved $1.14, $1.87 and $1.60 by using coupons. How much money did she give you this week?

This problem combines the money amounts. This is an addition problem.

$1.14
 1.87
+1.60

| save 10¢ | | save 25¢ |

You can use base ten blocks to add.

Exchange 10 hundredths for 1 tenth. Exchange 10 tenths for 1 whole.

Record.

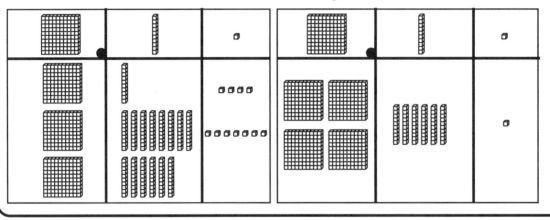

$1.14
 1.87
 1.60
$4.61

Find the sums and differences. Use base ten blocks.

1.   0.32
   + 0.23

2.   0.72
   + 0.86

3.   3.42
   + 4.25

4.   6.74
   + 1.32

5.   0.32
   − 0.21

6.   1.58
   − 0.72

7.   7.67
   − 4.25

8.   7.06
   − 6.78

9. 0.86 − 0.14 = _____

10. 0.52 − 0.18 = _____

11. 0.25 + 0.94 = _____

12. 0.44 + 0.56 = _____

Subtracting Decimal Fractions

You can use fraction bars, base ten blocks or a rule to subtract decimals.

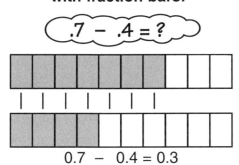

with fraction bars:	**with base ten blocks:** Build the greater number; remove the lesser number.

.7 − .4 = ?

0.7 − 0.4 = 0.3

$$\begin{array}{r} 2.65 \\ -\ 1.43 \\ \hline 1.22 \end{array}$$

ones	tenths	hundredths

with a rule: Line up the decimal points to subtract digits having the same place values. The decimal point in the difference is written in line with the other decimal points.

line up decimal points

0.9 − 0.48 = ?

$$\begin{array}{r} .90 \\ -\ .48 \\ \hline .42 \end{array}$$

Add 0's to hold place.

Find the sums and differences.

1.
$$\begin{array}{r} 0.9 \\ -\ 0.2 \\ \hline \end{array}$$

2.
$$\begin{array}{r} 5.4 \\ -\ 3.2 \\ \hline \end{array}$$

3.
$$\begin{array}{r} 4.3 \\ -\ 0.7 \\ \hline \end{array}$$

4.
$$\begin{array}{r} 2.8 \\ -\ 1.9 \\ \hline \end{array}$$

5.
$$\begin{array}{r} 3.45 \\ -\ 2.14 \\ \hline \end{array}$$

6.
$$\begin{array}{r} 4.5 \\ -\ 2.57 \\ \hline \end{array}$$

7.
$$\begin{array}{r} 7.79 \\ -\ 5.09 \\ \hline \end{array}$$

8.
$$\begin{array}{r} 20.4 \\ -\ 10.31 \\ \hline \end{array}$$

9.
$$\begin{array}{r} 8.07 \\ -\ 2.062 \\ \hline \end{array}$$

10.
$$\begin{array}{r} \$84.25 \\ -\ 9.75 \\ \hline \end{array}$$

11.
$$\begin{array}{r} \$0.74 \\ -\ 0.37 \\ \hline \end{array}$$

12.
$$\begin{array}{r} 476.13 \\ -\ 85.74 \\ \hline \end{array}$$

13. 0.84 − 0.39 = _____

14. 0.837 − 0.325 = _____

15. 3.6 − 0.24 = _____

16. 6 − 0.09 = _____

17. At the beginning of a trip, the odometer read 38152.9. At the end of the trip it read 38427.3. How long was the trip?

_____

18. Jan's height is 1.327 meters. Susan's height is 1.285 meters. How much taller is Jan than Susan?

_____

Multiplying Decimals: Tenths by Tenths

Sally saves one-tenth of the money she earns. If Sally earns 10¢, she will save $\frac{1}{10}$ of 10¢ or $\frac{1}{10}$ of one dime. How much will she save?

You can use base ten blocks to find one-tenth of one-tenth.
First, build one-tenth.
Then, exchange one-tenth for 10 hundredths.

$\frac{1}{10}$ of =

In words: one-tenth of one-tenth = one-hundredth.
In fractions: $\frac{1}{10}$ of $\frac{1}{10}$ = $\frac{1}{10} \times \frac{1}{10}$ = $\frac{1}{100}$
In decimals: $0.1 \times 0.1 = 0.01$

**Use base ten blocks to find each product. Draw a picture of your answer.
Write the number in words, in fractions and in decimals.**

1. $\frac{3}{10}$ of ▯ = _____	3 tenths of 1 tenth = 3 hundredths	$\frac{3}{10} \times \frac{1}{10} = \frac{3}{100}$	$0.3 \times 0.1 = 0.03$
2. $\frac{6}{10}$ of ▯ = _____			
3. $\frac{1}{10}$ of ▯ = _____			
4. $\frac{3}{10}$ of ▯ = _____			

5. Complete the table.

x	0.1	0.2	0.3	0.4	0.5	0.6	0.7	0.8	0.9
0.1									
0.2									
0.3									
0.4									
0.5									
0.6									
0.7									
0.8									
0.9									

6. Complete the patterns:

A one-tenth of one-tenth is

_____

B the word "of" means to _____

C 0.1 of 0.1 is _____

D the product of tenths (a <u>one</u>-place decimal) times tenths (a <u>one</u>-place decimal) is _____
(a _____ place decimal).

A Pattern for Multiplying Decimals

Rewrite a decimal problem as fractions to find the pattern.

$$\begin{array}{r} 0.6 \\ \times\ 0.3 \\ \hline \end{array} \qquad\qquad \begin{array}{r} 0.24 \\ \times\ 0.4 \\ \hline \end{array} \qquad\qquad \begin{array}{r} 0.09 \\ \times\ 0.04 \\ \hline \end{array}$$

$$\frac{6}{10} \times \frac{3}{10} = \frac{18}{100} \qquad \frac{4}{10} \times \frac{24}{100} = \frac{96}{1000} \qquad \frac{9}{100} \times \frac{4}{100} = \frac{36}{10000}$$

$$0.6 \times 0.3 = 0.18 \qquad\quad 0.4 \times 0.24 = 0.096 \qquad\quad 0.09 \times 0.04 = 0.0036$$

The pattern: Multiply as with whole numbers. The product will have the same number of decimal places as the total decimal places of the factors.

Find the products.

1.
$$\begin{array}{r} 3.25 \\ \times\ \ \ 3 \\ \hline \end{array}$$
_____ decimal places
_____ decimal places
_____ decimal places

2.
$$\begin{array}{r} 3.2 \\ \times\ 0.8 \\ \hline \end{array}$$
_____ decimal places
_____ decimal places
_____ decimal places

3.
$$\begin{array}{r} 29.8 \\ \times\ 0.06 \\ \hline \end{array}$$
_____ decimal places
_____ decimal places
_____ decimal places

4.
$$\begin{array}{r} 0.32 \\ \times\ 0.06 \\ \hline \end{array}$$
_____ decimal places
_____ decimal places
_____ decimal places

5.
$$\begin{array}{r} 0.5 \\ \times\ 0.3 \\ \hline \end{array}$$

6.
$$\begin{array}{r} 0.23 \\ \times\ 0.4 \\ \hline \end{array}$$

7.
$$\begin{array}{r} 0.27 \\ \times\ 0.04 \\ \hline \end{array}$$

8.
$$\begin{array}{r} 0.19 \\ \times\ \ \ 6 \\ \hline \end{array}$$

9.
$$\begin{array}{r} 0.014 \\ \times\ 0.4 \\ \hline \end{array}$$

10.
$$\begin{array}{r} 0.17 \\ \times\ 0.07 \\ \hline \end{array}$$

11.
$$\begin{array}{r} 4.9 \\ \times\ \ 5 \\ \hline \end{array}$$

12.
$$\begin{array}{r} 2.7 \\ \times\ 0.6 \\ \hline \end{array}$$

13.
$$\begin{array}{r} 3.5 \\ \times 0.04 \\ \hline \end{array}$$

14.
$$\begin{array}{r} 65.1 \\ \times 0.03 \\ \hline \end{array}$$

15.
$$\begin{array}{r} 2.3 \\ \times\ 0.6 \\ \hline \end{array}$$

16.
$$\begin{array}{r} 59.8 \\ \times 0.05 \\ \hline \end{array}$$

17. Gina's mother walks 2.75 kilometers each day. How many kilometers does she walk in one week?

_____

18. Your home uses 8.4 kilowatts of electricity each day. How much would be used in 0.5 of a day?

_____

Dividing Decimals by Whole Numbers: Find the Pattern

You can find the pattern (or steps) in division as you share the blocks.

$$3\overline{)4.62}$$

$$\begin{array}{r} 1 \\ 3\overline{)4.62} \\ 3 \\ \hline 1\,6 \end{array}$$

← **D**ivide.

← **M**ultiply.

← **S**ubtract and ↓ **bring down** the next number.

Repeat the **DMS** ↓ pattern with the tenths blocks.

$$\begin{array}{r} 1.5 \\ 3\overline{)4.62} \\ 3 \\ \hline 1\,6 \\ 1\,5 \\ \hline 1\,2 \end{array}$$

← D

← M

← S↓

Repeat the **DMS** ↓ pattern with the hundredths blocks.

$$\begin{array}{r} 1.54 \\ 3\overline{)4.62} \\ 3 \\ \hline 1\,6 \\ 1\,5 \\ \hline 1\,2 \\ 1\,2 \\ \hline \end{array}$$

← D

← M

Find the quotients.

1. $2\overline{)2.84}$

2. $7\overline{)27.23}$

3. $5\overline{)183.5}$

4. $6\overline{)193.2}$

5. $3\overline{)10.26}$

6. $8\overline{)19.28}$

7. $8\overline{)58.48}$

8. $9\overline{)6.282}$

9. $6\overline{)\$32.22}$

10. $9\overline{)\$1.17}$

11. $8\overline{)\$32.80}$

12. $5\overline{)\$23.05}$

13. Four bottles hold 5.68 liters of tomato juice. If there are equal amounts in each bottle, how much tomato juice is in one bottle?

14. Jesús and his father canned 4.5 liters of salsa. They put the same amount in each of 6 jars. How much salsa was put in each jar?

Dividing by Decimals

A box of mixed nuts weighs 1 pound. The nuts are divided into smaller bags weighing 0.2 pounds each. How many bags can be filled?

$1 \div 0.2$

$0.2\overline{)1.0}$

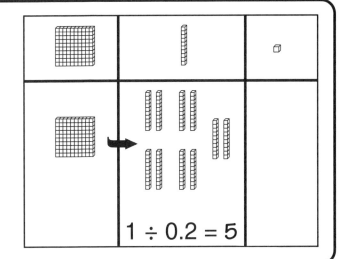

$1 \div 0.2 = 5$

Use base ten blocks to find the quotients.

1. $0.1\overline{)0.3}$ **2.** $0.3\overline{)0.6}$ **3.** $0.2\overline{)0.8}$ **4.** $0.5\overline{)1}$

5. $0.01\overline{)0.04}$ **6.** $0.02\overline{)0.06}$ **7.** $0.05\overline{)1}$ **8.** $0.04\overline{)1}$

Rewrite each problem as an equivalent problem with a whole number divisor.
Hint: Multiply the divisor and dividend by 10 or 100.
Find the quotients.

9. $0.3\overline{)0.9}$ becomes $3\overline{)9}$ **10.** $0.5\overline{)1}$ becomes $\overline{)}$

11. $0.4\overline{)3.56}$ becomes $\overline{)}$ **12.** $0.5\overline{).625}$ becomes $\overline{)}$

13. $0.03\overline{)1.74}$ becomes $\overline{)}$ **14.** $0.02\overline{)4.76}$ becomes $\overline{)}$

15. $0.06\overline{)1.506}$ becomes $\overline{)}$ **16.** $0.06\overline{).018}$ becomes $\overline{)}$

Positive and Negative Numbers on a Number Line

A thermometer turned sideways becomes a number line showing positive and negative numbers.

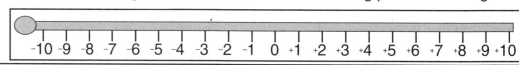

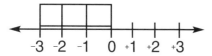

The white squares can be thought of as being to the left of the zero on the number line.

$$\square\ \square\ \square\ =\ -3$$

The black squares can be thought of as being to the right of zero on the number line.

$$\blacksquare\ \blacksquare\ \blacksquare\ =\ +3$$

Draw black or white squares on the number line to show each number.

1. +5

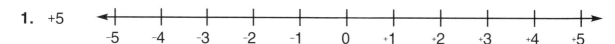

2. −4

How many black or white squares from zero to each point on the number line?

3.

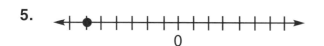

_____

4.

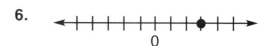

_____

5.

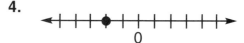

_____

6.

_____

7. The temperature at noon was 10° above zero. By 10 p.m., the temperature had dropped 15°. What was the temperature at 10 p.m.?

_____

8. The temperature at midnight was 10° below zero. At noon the next day, the temperature was 20° above zero. How many degrees had the temperature increased?

_____

Opposites Represent Zero

+2 and −2 cancel each other out.	+3 and −2 do not cancel each other out.
■ ■ □ □	+3 and −2 are not opposites.
+2 + (−2) represents the number zero.	■ ■ ■ □ □
	+3 +(−2) does not represent the number zero.

Write the integer names for the number of black and white cubes in each set. Indicate (yes or no) if the set represents zero.

1. □ □ □
    ■ ■ ■   _____ and _____
                 yes   or   no

2. ■ ■ ■
    ■ ■ □   _____ and _____
                 yes   or   no

3. □   ■
    □ ■ ■   _____ and __
                 yes   or   no

4. ■ ■
    □ □   _____ and _____
                 yes   or   no

5. □ □ ■
    □ □ ■   _____ and _____
                 yes   or   no

6. □ □ ■ ■
    ■ ■ □ □   _____ and _____
                 yes   or   no

Name the opposite of each number.

7. −5 _____

8. 2 _____

9. 12 _____

10. −3 _____

11. 0 _____

12. 6 _____

13. 54 _____

14. −72 _____

True or False

_____ 15. Every positive number has an opposite.

_____ 16. Every negative number has an opposite.

_____ 17. Combining any number with its opposite results in +1.

_____ 18. Zero is greater than any negative integer.

_____ 19. Every positive integer is greater than every negative integer.

_____ 20. Combining any number with its opposite results in −1.

Adding Positive and Negative Integers

Cathy gained 2 pounds on the weekend. She lost 5 pounds during the week. What was the net change in her weight? 	Use two black cubes to represent gains and five white cubes to represent losses. 
Join the two numbers. One positive cube cancels out one negative cube. How many cubes are unmatched? What color are the unmatched cubes? 	There are three unmatched white cubes. $2 + (-5) = -3$

Write the integers. Count to find the sum.

1.

_____ + _____ = _____

2.

_____ + _____ = _____

3.

_____ + _____ = _____

4.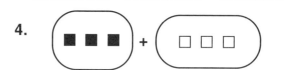

_____ + _____ = _____

5.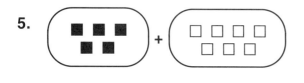

_____ + _____ = _____

6.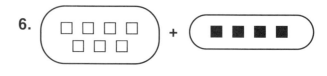

_____ + _____ = _____

Which of your answers were positive?
Can you find the pattern for the sign of your answer for
adding a positive and negative number?

Adding Negative Integers

A football team lost 5 yards on the first play and 4 yards on the second play. What was the net gain or loss?

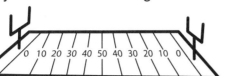

Use white cubes to show addition of negative integers.

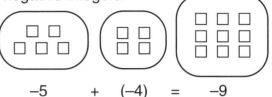

$-5 \quad + \quad (-4) \quad = \quad -9$

Write the integers. Count to find the sum.

1. ⬜⬜⬜ + ⬜⬜⬜⬜

_____ + _____ = _____

2. ⬜⬜⬜ + ⬜⬜⬜⬜⬜

_____ + _____ = _____

3. ⬜⬜⬜ + ⬜⬜

_____ + _____ = _____

4. ⬜⬜⬜⬜ + ⬜⬜⬜⬜

_____ + _____ = _____

Draw pictures of each sum.

5. $-2 + (-3) =$

6. $-4 + (-3) =$

7. $-3 + (0) =$

8. $-3 + (-3) =$

9. $-10 + (-5) =$

10. $-6 + (-12) =$

11. $-8 + 0 =$

12. $-30 + (-50) =$

13. $-74 + (-7) =$

14. $-32 + (-45) =$

15. $-37 + (-8) =$

16. $-123 + (-304) =$

17. $-536 + (-245) =$

18. $-240 + (-385) =$

19. $-206 + (-347) =$

20. $-458 + (-678) =$

How many of your answers were negative?
Can you find the pattern for the sum of two
negative integers?

Subtracting Negative Integers

You can use black and white cubes to subtract negative integers. You may have to add pairs of black and white cubes (each pair is a sum of zero) to have enough cubes to subtract.

−2 − (−3)

Build −2 with 2 white cubes. There aren't enough to subtract...

Add 1 white cube and 1 black cube so you can subtract.

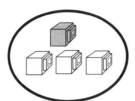

Subtract 3 white cubes. There is 1 black cube left.

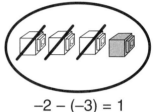

−2 − (−3) = 1

1. −4 − (−1) = _____

2. −3 − (−2) = _____

3. −2 − (−1) = _____

4. −3 − (−4) = _____

5. −2 − (−4) = _____

6. −1 − (−3) = _____

7. 2 − (−1) = _____

8. 3 − (−2) = _____

9. 2 − (−2) = _____

10. −3 − 2 = _____

11. −2 − (−3) = _____

12. −4 − (−2) = _____

Draw pairs of black and white cubes in each set so that you can subtract.

13. −1 − (−4) = _____

14. 1 − (−2) = _____

15. −2 − 1 = _____

Can you find the pattern for subtracting with negative integers?

Adding Opposites: a Shortcut for Subtraction

Subtracting a number gives the same result as adding its opposite.

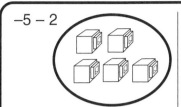

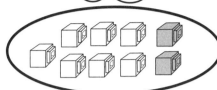

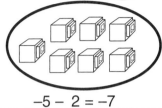

$-5 - 2$

There aren't enough black cubes.

Add 2 black and 2 white cubes so you can subtract.

Subtract 2 black cubes. There are 7 white cubes left.

$-5 - 2 = -7$

Shortcut: Add the opposite of the number being subtracted.

$-5 - 2$ The opposite of 2 is –2 $-5 + (-2) = -7$

Write the opposite of each number.

1. 7 _____

2. –5 _____

3. 2 _____

4. –4 _____

5. 0 _____

6. $\frac{3}{4}$ _____

7. $-\frac{1}{3}$ _____

8. –2.5 _____

Rewrite each subtraction problem as an addition of opposites. Solve.

9. $4 - (-2)$

  $\underline{\ 4 + 2\ }$ = $\underline{\ 6\ }$

10. $-2 - 3$

  _____ = ___

11. $-5 - (-2)$

  _____ = ___

12. $-4 - 6$

  _____ = ___

13. $5 - (-1)$

  _____ = ___

14. $-3 - 4$

  _____ = ___

15. $-6 - 6$

  _____ = ___

16. $5 - 0$

  _____ = ___

17. $-4 - 2$

  _____ = ___

18. $-3 - (-1)$

  _____ = ___

19. $-4 - (-5)$

  _____ = ___

20. $-12 - 5$

  _____ = ___

21. $2 - 4$

  _____ = ___

22. $0 - (-1)$

  _____ = ___

23. $5 - (-5)$

  _____ = ___

Multiplying by Positive Integers

You can use a number line or black and white cubes to multiply by a positive integer.

positive x positive	positive x negative
Bruce saved $5 a week for 3 weeks. How much did he save?	The average weight loss at the diet clinic is 2 pounds per week. How many pounds can Maria expect to lose in 4 weeks?

positive x positive

Start at 0. Take 3 jumps of 5 to the right.

$5 + 5 + 5$ or $3 \times 5 = 15$

Put together 3 groups of 5 black cubes each.

$3 \times 5 = 15$

positive x negative

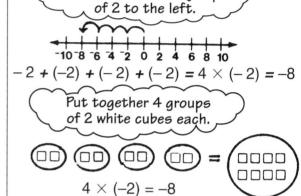

Start at 0. Take 4 jumps of 2 to the left.

$-2 + (-2) + (-2) + (-2) = 4 \times (-2) = -8$

Put together 4 groups of 2 white cubes each.

$4 \times (-2) = -8$

Show jumps on the number line to solve each problem.

1. $4 \times 2 =$ _____

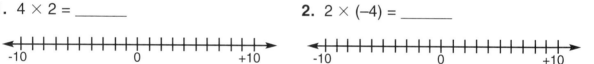
-10   0   +10

2. $2 \times (-4) =$ _____

-10   0   +10

3. $3 \times (-2) =$ _____

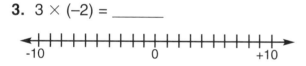
-10   0   +10

4. $3 \times 0 =$ _____

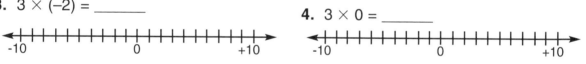
-10   0   +10

Write the multiplication example for each problem. Use black and white cubes to multiply.

5. Put together 2 groups of 5.

_____ × _____ = _____

6. Put together 3 groups of –2.

_____ × _____ = _____

7. Put together 3 groups of –3.

_____ × _____ = _____

8. Put together 4 groups of 2.

_____ × _____ = _____

Can you find a pattern for multiplying a positive integer by a positive integer?

Can you find a pattern for multiplying a positive integer by a negative integer?

Multiplying by Negative Integers

To multiply by a positive integer, you can picture taking black cubes out of a set.
To multiply by a negative integer, you can picture taking white cubes out of a set.

negative x positive	negative x negative
$-3 \times 2 = ?$	$-2 \times (-4) = ?$
1. Start with a set of black and white cubes representing zero. (There must be the same number of black and white cubes).	1. Start with a set of zero.
2. Remove 3 sets of 2 black cubes each. How many are left?	2. Remove 2 sets of 4 white cubes each. How many are left?
There are 6 unmatched white cubes left.	*There are 8 unmatched black cubes left.*
$(-3) \times 2 = -6$	$-2 \times (-4) = {}^{+}8$

Write the multiplication example for each problem. Use black and white cubes to solve each problem.

1. Take away 3 groups of 2
   _____ × _____ = _____

2. Take away 2 groups of –3
   _____ × _____ = _____

3. Take away 2 groups of –4
   _____ × _____ = _____

4. Take away 3 groups of 3
   _____ × _____ = _____

5. $-7 \times (-3) =$ ___

6. $-6 \times 4 =$ ____

7. $-8 \times 4 =$ ____

8. $-9 \times 3 =$ _____

9. $-8 \times (-5) =$ ___

10. $-7 \times (-2) =$ ____

11. $-9 \times (-6) =$ ____

12. $-6 \times (-6) =$ ____

13. $-6 \times 8 =$ ____

14. $-9 \times 9 =$ ____

15. $-7 \times (-7) =$ ____

16. $-8 \times (-9) =$ ____

17. $-1 \times 10 =$ ____

18. $-2 \times (-21) =$ ___

19. $-4 \times 0 =$ _____

20. $-5 \times (-13) =$ ___

Can you find a pattern for multiplying a negative integer by a negative integer?

Dividing Integers with the Same Sign

You can find the pattern for the quotient by taking away the same size group or by relating division to multiplication.

<u>taking away groups</u>   <u>relating to multiplication</u>

$6 \div 2$ → 3 groups

$^-6 \div (-2)$ → 3 groups

If $6 \div \Box = 2$, $2 \times \Box = 6$

$\Box = 3$

If $-6 \div (-2) = \Box$, $-2 \times \Box = -6$

$\Box = 3$

Use black and white cubes to find the quotients.

1. $10 \div 2$ _____ **2.** $-8 \div (-4)$ _____ **3.** $9 \div 3$ _____ **4.** $-6 \div (-3)$ _____

5. $-8 \div (-2)$ _____ **6.** $4 \div 2$ _____ **7.** $-6 \div (-2)$ _____ **8.** $5 \div 5$ _____

Write the multiplication sentence for each division sentence. Solve for $\Box$.

9. $15 \div 5 = \Box$ **10.** $18 \div 9 = \Box$ **11.** $-28 \div (-7) = \Box$ **12.** $-32 \div (-4) = \Box$

$\underline{5} \times \Box = \underline{15}$ $\underline{9} \times \Box = \underline{18}$ $\_\_ \times \_\_ = \_\_$ $\_\_ \times \_\_ = \_\_$

$\Box = \_\_$ $\Box = \_\_$ $\Box = \_\_$ $\Box = \_\_$

13. $-36 \div (-6) = \Box$ **14.** $56 \div 7 = \Box$ **15.** $^-64 \div (-8) = \Box$ **16.** $-48 \div (-8) = \Box$

$\_\_ \times \_\_ = \_\_$ $\_\_ \times \_\_ = \_\_$ $\_\_ \times \_\_ = \_\_$ $\_\_ \times \_\_ = \_\_$

$\Box = \_\_$ $\Box = \_\_$ $\Box = \_\_$ $\Box = \_\_$

17. $-49 \div (-7) = \Box$ **18.** $42 \div 6 = \Box$ **19.** $-54 \div (-9) = \Box$ **20.** $-63 \div (-7) = \Box$

$\_\_ \times \_\_ = \_\_$ $\_\_ \times \_\_ = \_\_$ $\_\_ \times \_\_ = \_\_$ $\_\_ \times \_\_ = \_\_$

$\Box = \_\_$ $\Box = \_\_$ $\Box = \_\_$ $\Box = \_\_$

Can you find the pattern for the sign of the quotient when dividing integers with the same signs?

Dividing Integers

The patterns for dividing integers can be found by relating division to multiplication.

Multiplication	Related Division	Division Patterns
$4 \times 2 = 8$	$8 \div 4 = 2$	positive ÷ positive = positive
$-4 \times (-2) = 8$	$8 \div (-2) = -4$	positive ÷ negative = negative
$-4 \times 2 = -8$	$-8 \div 2 = -4$	negative ÷ positive = negative
$4 \times (-2) = -8$	$-8 \div (-2) = 4$	negative ÷ negative = positive

Write N for negative and P for positive for the answer.

1. $P \times N = $ ___

2. $P \times P = $ ____

3. $N \times N = $ ____

4. $N \times P = $ ____

5. $P \div P = $ ____

6. $P \div N = $ ____

7. $N \div P = $ ____

8. $N \div N = $ ____

9. $-42 \div 6 = $ ____

10. $-18 \div (-6) = $ ___

11. $63 \div (-7) = $ ____

12. $0 \div (-2) = $ ____

13. $-48 \div (-8) = $ ___

14. $35 \div (-7) = $ ____

15. $-36 \div 6 = $ ____

16. $40 \div (-8) = $ ____

17. $72 \div (-8) = $ ____

18. $-28 \div (-7) = $ ___

19. $32 \div (-4) = $ ____

20. $-30 \div (-5) = $ ___

21. $12 \div 4 = $ ____

22. $-9 \div (-3) = $ ____

23. $14 \div (-2) = $ ____

24. $-7 \div 7 = $ ____

25. $54 \div (-9) = $ ____

26. $-16 \div 4 = $ ___

27. $9 \div (-9) = $ ____

28. $0 \div 8 = $ ____

29. $-60 \div (-10) = $ ___

30. $44 \div (-11) = $ ____

31. $-36 \div 12 = $ ___

32. $120 \div -20 = $ ___

33. The quotient of two numbers with the same sign is always _____.

34. The quotient of two numbers with different signs is always _____.

Taking Rational Numbers to Powers

Stacy is making cookies, but she only wants to make $\frac{3}{4}$ of a batch.
If the recipe calls for $\frac{3}{4}$ cup of butter, how much butter does she need?

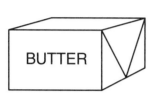
BUTTER

$$\frac{3}{4} \text{ of } \frac{3}{4} \text{ cup} = \frac{3}{4} \times \frac{3}{4} \text{ cup} = \frac{3 \times 3}{4 \times 4} \text{ cup} = \frac{9}{16} \text{ cup}$$

We can write this answer in exponential form.

$$\frac{3}{4} \times \frac{3}{4} = \frac{3 \times 3}{4 \times 4} = \frac{3^2}{4^2}$$

Another way to write $\frac{3^2}{4^2}$ is $\left(\frac{3}{4}\right)^2$

Write the following exponents in expanded form and find the value of the expression.

1. $\left(\frac{2}{3}\right)^4 = \frac{2^4}{3^4} = \frac{2 \times 2 \times 2 \times 2}{3 \times 3 \times 3 \times 3} = \frac{16}{81}$ 2. $\left(\frac{7}{9}\right)^3 =$

3. $\left(\frac{1}{4}\right)^2 =$ 4. $\left(\frac{3}{5}\right)^3 =$

5. $(0.3)^2 = 0.3 \times 0.3 = 0.09$ 6. $(1.7)^4 =$

Write the following expressions using one exponent.

7. $\frac{5 \times 5 \times 5 \times 5}{6 \times 6 \times 6 \times 6} =$ 8. $\frac{12 \times 12}{17 \times 17} =$

9. $\frac{3 \times 3 \times 3 \times 3 \times 3 \times 3}{5 \times 5 \times 5 \times 5 \times 5 \times 5} =$ 10. $\frac{4 \times 4 \times 4 \times 4 \times 4 \times 4 \times 4 \times 4 \times 4}{9 \times 9 \times 9 \times 9 \times 9 \times 9 \times 9 \times 9 \times 9} =$

11. $2.4 \times 2.4 \times 2.4 \times 2.4 =$ 12. $0.64 \times 0.64 \times 0.64 =$

Interchanging Decimals and Fractions

When the denominator of a fraction is 10 or 100, the denominator tells the place value where the last digit will be written.

Change $\frac{21}{100}$ to a decimal.

$$0 \, . \, \underline{2} \, \underline{1}$$
$$\text{hundredths}$$

The place value of the last digit in a decimal indicates the denominator of the fraction.

Change 1.23 to a fraction.

$$1 \, . \, 2 \, \underline{3} \qquad = 1\frac{23}{100}$$
$$\text{hundredths}$$

Set A. Change each fraction to a decimal.

1. $\frac{5}{10}$ _____

2. $\frac{3}{10}$ _____

3. $\frac{2}{10}$ _____

4. $\frac{9}{10}$ _____

5. $1\frac{7}{10}$ _____

6. $1\frac{1}{10}$ _____

7. $2\frac{4}{10}$ _____

8. $2\frac{9}{10}$ _____

9. $\frac{13}{100}$ _____

10. $\frac{41}{100}$ _____

11. $\frac{82}{100}$ _____

12. $\frac{51}{100}$ _____

13. $1\frac{27}{100}$ _____

14. $4\frac{15}{100}$ _____

15. $6\frac{37}{100}$ _____

16. $8\frac{92}{100}$ _____

17. $\frac{10}{10}$ _____

18. $\frac{20}{10}$ _____

19. $\frac{5}{100}$ _____

20. $\frac{7}{100}$ _____

21. $\frac{4}{10}$ _____

22. $\frac{17}{100}$ _____

23. $1\frac{3}{10}$ _____

24. $2\frac{59}{100}$ _____

Set B. Change each decimal to a fraction.

1. 0.6 _____

2. 0.7 _____

3. 0.9 _____

4. 0.3 _____

5. 0.21 _____

6. 0.69 _____

7. 0.37 _____

8. 0.83 _____

9. 1.7 _____

10. 1.1 _____

11. 2.4 _____

12. 2.9 _____

13. 1.27 _____

14. 4.17 _____

15. 6.37 _____

16. 8.91 _____

17. 1.0 _____

18. 0.03 _____

19. 0.09 _____

20. 1.07 _____

Percent: Another Name for Hundredths

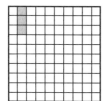

$\frac{1}{100}$

The shaded part is $\frac{1}{100}$ or 0.01 of the whole.

Percent is another name for "parts per hundred" or hundredths.

The shaded part can also be written as 1 percent or 1%.

What part is shaded? Write your answer as a fraction, as a decimal and as a percent.

1.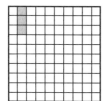

2.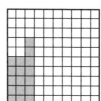

3.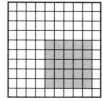

_____ _____ _____   _____ _____ _____   _____ _____ _____

4.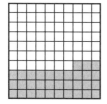

5.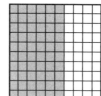

6.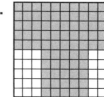

_____ _____ _____   _____ _____ _____   _____ _____ _____

7.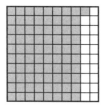

8.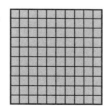

9.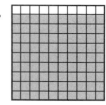

_____ _____ _____   _____ _____ _____   _____ _____ _____

Changing Fractions to Percents Using Models

The large square is divided into 100 small squares. Each small square is

$\frac{1}{100}$ or 1%

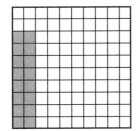

16 of the 100 small squares are shaded.
$\frac{16}{100}$ or 16% is shaded.

To change $\frac{1}{5}$ to a percent, you can think of how many small squares would be shaded if $\frac{1}{5}$ of the large square is shaded.

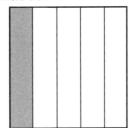 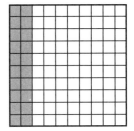

$\frac{1}{5}$ of the square is the same as 20%

Shade in the fractional part of each square.
Tell what percent is shaded by the fractional part.

1.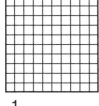

$\frac{1}{2} = $ _____ %

2.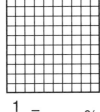

$\frac{1}{4} = $ _____ %

3.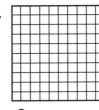

$\frac{3}{4} = $ _____ %

4.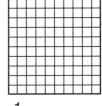

$\frac{1}{1} = $ _____ %

5.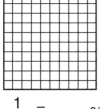

$\frac{1}{10} = $ _____ %

6.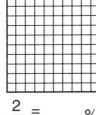

$\frac{2}{5} = $ _____ %

7.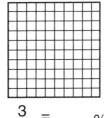

$\frac{3}{10} = $ _____ %

8.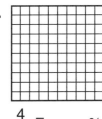

$\frac{4}{5} = $ _____ %

9.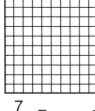

$\frac{7}{10} = $ _____ %

10.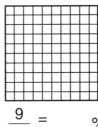

$\frac{9}{10} = $ _____ %

11.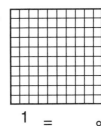

$\frac{1}{50} = $ _____ %

12.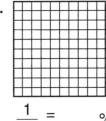

$\frac{1}{25} = $ _____ %

Mixed Practice

Write as a decimal.

1. 35% _____   2. 94% _____   3. 25% _____   4. 31% _____

5. 7% _____   6. 8% _____   7. 5% _____   8. 1% _____

Write as a percent.

9. $\frac{37}{100}$ _____   10. $\frac{9}{10}$ _____   11. $\frac{1}{4}$ _____   12. $\frac{1}{5}$ _____

13. $\frac{1}{2}$ _____   14. $\frac{6}{50}$ _____   15. $\frac{8}{25}$ _____   16. $\frac{3}{20}$ _____

Write as a percent.

17. 0.39 _____   18. 0.50 _____   19. 0.25 _____   20. 0.46 _____

21. 0.05 _____   22. 0.09 _____   23. 0.7 _____   24. 0.3 _____

Write as a fraction. Simplify if possible.

25. 20% _____   26. 11% _____   27. 80% _____   28. 50% _____

29. 35% _____   30. 24% _____   31. 52% _____   32. 10% _____

33. A football field has 100 yards. Mark ran 6 yards. What percent of the field did he run?

_____

34. Bob ran 7 yards and 15 yards on a football field. What percent of the field did Bob run?

_____

More Than 100%

The 1790 population of Jamestown was 250% of what the population was in 1780. How many times larger was the 1790 population than the 1780 population?

You can think about the meaning of 250%:

$100\% = \frac{100}{100}$ or all of it

$200\% = \frac{200}{100}$ or 2 times it

$250\% = \frac{250}{100}$ or 2½ times it

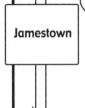

Jamestown

... or you can use the rule for changing a percent to a decimal.

$250\% = 250.\% = 250.\%$
$= 2.5$

Change each percent to a decimal.

1. 300% _____
2. 125% _____
3. 240% _____
4. 365% _____

5. 400% _____
6. 500% _____
7. 450% _____
8. 275% _____

Change each decimal to a percent.

9. 1.00 _____
10. 2.00 _____
11. 3.50 _____
12. 2.75 _____

13. 4.30 _____
14. 2.5 _____
15. 1.2 _____
16. 5.7 _____

One of the answers given for each problem is correct. Choose the correct answer.

17. Last year, Jack's brother weighed 40 lb. This year he weighs 80 lb. What percent of last year's weight is this year's weight?

    2%   20%   40%   200%

18. Karen started a stamp collection with 100 stamps. Now she has 350 stamps. How many times larger is her stamp collection now?

    3.5   35   250   350

19. The goal for the Ramsey School can collection was 1,000 cans. 3,000 cans were collected. What percent of the goal was collected?

    30%   100%   200%   300%

20. Nancy's goal was to swim 100 laps by the end of the summer. At the end of the summer, she could swim 50 laps. What percent of her goal had she acheived?

    50%   100%   150%   200%

Finding a Percent of a Number: Changing to Decimal

Once you have learned to estimate your answer with percent, it's time to find the actual answer. One of the most common ways is to change the percent to a decimal and multiply.

Jerry's math test paper was marked 72% correct. There were 50 problems on the test. How many problems did Jerry get correct?

$$72\% \text{ of } 50 = ?$$

Estimate: since Jerry got over half (50%) of the problems correct, the answer should be between 25 and 50.

Step 1. Change the percent to a decimal.

$$72\% = 0.72$$

Step 2. Multiply.

$$\begin{array}{r} 50 \\ \times\, 0.72 \\ \hline 100 \\ 350 \\ \hline 36.00 \end{array}$$

Move the decimal point two places.

36 problems correct

Change to decimals.

1. 34% _____
2. 63% _____
3. 87% _____
4. 29% _____

5. 18% _____
6. 35% _____
7. 6% _____
8. 5% _____

Multiply.

9. $\begin{array}{r} 43 \\ \times\ 0.32 \\ \hline \end{array}$

10. $\begin{array}{r} 67 \\ \times\ 0.05 \\ \hline \end{array}$

11. $\begin{array}{r} 2.60 \\ \times\ 0.45 \\ \hline \end{array}$

12. $\begin{array}{r} 32.50 \\ \times\ 0.04 \\ \hline \end{array}$

Estimate your answer. Then find the actual answer.

13. 24% of 52 = _____
    Estimate = _____

14. 46% of 23 = _____
    Estimate = _____

15. 19% of 46 = _____
    Estimate = _____

16. 28% of 25 = _____
    Estimate = _____

17. 12% of 79 = _____
    Estimate = _____

18. 6% of 52 = _____
    Estimate = _____

19. There are 30 students in a gym class. 40% of them are girls. How many are girls?

    Estimate: _____

    Actual: _____

20. There are 750 students in a school. 48% of them are boys. How many are boys?

    Estimate: _____

    Actual: _____

Comparing Two Numbers as a Percent

You already know how to compare two numbers as a fraction or ratio. You can compare two numbers as a percent by changing the fraction to a percent.

Example: There are 8 students. Two of the students wear glasses. What percent of the students wear glasses? Think: 2 is what percent of 8?

$2 =$ _____ % of 8

1. Write the numbers as a fraction.

Think: 2 is what part of 8?

$$\frac{2}{8} = \frac{1}{4}$$

2. Change the fraction to a percent.

$$\frac{1}{4} = 25\%$$

If you do not have this memorized, then divide 4 into 1.

$$4)\overline{100}^{.25} = 25\%$$

1. Complete the table. Compare the two numbers as a fraction and percent.

	a. 5, 10	b. 1, 3	c. 12, 16	d. 6, 10	e. 20, 30	f. 9, 36
fraction						
percent						

Fill in the blanks.

2. 12 is _____% of 20

3. 1 is _____% of 8

4. 35 is _____% of 40

5. 24 is _____% of 30

6. 4 is _____% of 16

7. 5 is _____% of 25

8. 6 is _____% of 18

9. 50 is _____% of 75

10. 40 is _____% of 400

11. There are 10 sandwiches. Six of them are cheese sandwiches. What percent are cheese sandwiches?

_____

12. There are 27 trees in a yard. Nine of the trees are Norway pine. What percent are Norway pine?

_____

13. There are 36 cupcakes. Twenty-four of them are chocolate. What percent are chocolate?

_____

14. There are 30 students. Ten of them are wearing tennis shoes. What percent are wearing tennis shoes?

_____

Finding the Percent of Increase or Decrease

You already know how to compare two numbers as a percent. You can use that skill to help you understand how much change occurs to a number over a period of time. We call this change the percent of increase or decrease.

A loaf of bread cost 30¢ in 1950 and 35¢ in 1955. What was the percent of increase in the cost of bread?

Think: Compare the amount of change to the original.

$$\% \text{ of change} = \frac{\text{change}}{\text{original}}$$

1. Write a ratio comparing the change to the original.
   change = 35¢ − 30¢ = 5¢
   original price = 30¢
   $$\frac{5}{30} = \frac{1}{6}$$

2. Change the ratio to a percent.
   $$\frac{1}{6} = 16\frac{2}{3}\%$$

There were 500 students enrolled in West Junior High last year and 450 students this year. What was the percent of decrease in the enrollment?

Don't forget to compare the amount it changed to the original.

$$\% \text{ of change} = \frac{\text{change}}{\text{original}}$$

1. Write a ratio comparing the change to the original.
   change = 500 − 450 = 50
   original = 500
   $$\frac{50}{500} = \frac{1}{10}$$

2. Change the ratio to a percent.
   $$\frac{1}{10} = 10\%$$

   10% decrease

Find the percent of increase or decrease for each problem.

		Price last week	Price this week	Change	Percent of Change
1.	Eggs	50¢	40¢	_____	_____
2.	Milk	22¢	25¢	_____	_____
3.	Dress	$24	$20	_____	_____
4.	Butter	80¢	88¢	_____	_____
5.	Candy Bar	12¢	15¢	_____	_____
6.	Paul's weight, formerly 80 lb., now 84 lb.			_____	_____
7.	Enrollment last year 500, now 480.			_____	_____
8.	Shirt was $7.40, now $6.00.			_____	_____
9.	Value of auto was $750, now $690.			_____	_____
10.	John's height was 60 in., now 65 in.			_____	_____
11.	Wages were $90, now $110.			_____	_____
12.	Auto value was $800, now $715.			_____	_____
13.	Home value was $30,000, now $50,000.			_____	_____
14.	Susan's weight was 80 lb., now 76 lb.			_____	_____
15.	Mark's weight was 50 kg, now 52 kg.			_____	_____

Percent of Increase Over 100%

A percent of increase can be over 100% if a number increases by a number larger than itself.

The population of Jonesville was 60,000 in 1970 and 150,000 in 1980.
What was the percent of increase?

% of change = $\dfrac{\text{change}}{\text{original}}$

1. Write a ratio comparing the change to the original.

2. Change the ratio to a percent.

change = 150,000 − 60,000 = 90,000
original = 60,000

$\dfrac{90,000}{60,000} = \dfrac{3}{2}$

$2 \overline{)3.00} = 150\%$ with 1.50 on top

150% increase in population

Find the percent of increase.

1. From 20° to 40°.

2. From $36 to $45.

3. From $5,000 to $12,000.

4. From $6,000 to $15,000.

5. The Boy Scouts sold 150 wreaths for Christmas this year and 180 last year. Find the percent of change.

6. Harvey now weighs 80 pounds. A year ago he weighed 75 pounds. What percent has his weight increased?

7. The temperature at 6 a.m. was 10°. At noon it was 30°. What was the percent of increase in temperature?

8. The sales in a record shop were $120,000 last year and $140,000 this year. What was the percent of increase?

9. The value of a car depreciated from $4400 to $3300. Find the percent of depreciation.

10. The value of a home went from $50,000 to $70,000. What was the percent of appreciation (increase in value)?

11. A used auto cost $1200. After one year, it had depreciated $400. What was the percent of depreciation?

12. A dress marked $48.00 was on sale for $40.00. What was the rate of decrease?

Find the percent of increase when:
A) a number increases by itself.
B) a number increases by twice itself.
C) a number increases by 1½ times itself.

Finding the Net Price or Sale Price

Discount is a percent of savings that is subtracted from the regular price. The amount paid (after the discount has been subtracted) is called the net price.

A sweater costs $36.00. It is offered on sale for 25% off.
Find the sale price (the net price).

1. Find the discount by changing the percent to a fraction or a decimal.

$\frac{1}{4} \times \$36.00 = \$9.00$
or
$.25 \times \$36 = \$9.00$

The discount is $9.00

2. Subtract the discount to find the net price.

$36.00
− 9.00
$27.00

The net price (sale price) is $27.00.

1. A record sells regularly for $8.95. It is on sale for 40% off. Find the discount amount and the sale price.

$ _____ $ _____
Discount   Sale Price

2. Ski gloves priced at $9 are on sale for $33\frac{1}{3}$% discount. What is the discount amount and the sale price?

$ _____ $ _____
Discount   Sale Price

Find the discount and the net price for each item below. Use either decimals or fractions.

3. Bicycle
Regular price = $110
Discount = 10%

$ _____ $ _____
Discount   Sale Price

4. Jeans
Regular price = $28
Discount = $12\frac{1}{2}$%

$ _____ $ _____
Discount   Sale Price

5. Down Vest
Regular price = $45.00
Discount = 50%

$ _____ $ _____
Discount   Sale Price

6. Downhill Skis
Regular price = $336.00
Discount = $66\frac{2}{3}$%

$ _____ $ _____
Discount   Sale Price

7. VCR
Regular price = $789.00
Discount = 20%

$ _____ $ _____
Discount   Sale Price

8. Computer Game
Regular price = $69.50
Discount = 30%

$ _____ $ _____
Discount   Sale Price

9. Scarf
Regular price = $15
Discount = 70%

$ _____ $ _____
Discount   Sale Price

10. Ski Boots
Regular price = $68.00
Discount = 25%

$ _____ $ _____
Discount   Sale Price

Finding Interest and Amount Due: Part of a Year

Jim wanted to buy a motorcycle. He was short $550. He borrowed the money for 6 months at 12%. How much interest did he pay? What was the total amount he paid back?

| To find the interest:
$I = prt$
$I = \$550 \times 12\% \times 6$ months
$I = \$550 \times .12 \times \frac{1}{2}$ 6 months is $\frac{1}{2}$ year.
① Multiply the first two numbers.
$\begin{array}{r} 550 \\ \times \quad .12 \\ \hline 1100 \\ 550 \\ \hline \$66.00 \end{array}$ | ② Multiply the answer by the third number.
$$\$66 \times \frac{1}{2} = \$33$$

To find the amount due:
Add the principal plus the interest.
principal → $550.00
interest →   + $33.00
amount due → $583.00 |

Complete the chart to find the interest and the amount due.

	Amount Borrowed (Principal)	Rate	Time	Interest	Amount Repaid
1.	$800	8%	1 yr		
2.	$800	8%	6 mo		
3.	$800	14%	1 yr		
4.	$1500	11%	2 yr		
5.	$60,000	8%	30 yr		
6.	$60,000	12%	30 yr		
7.	$60,000	12%	25 yr		
8.	$175	10%	3 mo		
9.	$250	7%	4 mo		
10.	$4500	9%	$2\frac{1}{2}$ yr		
11.	$2000	13%	$3\frac{1}{2}$ yr		
12.	$1800	5%	4 yr		
13.	$2600	6%	6 mo		
14.	$75	8%	3 mo		
15.	$10,000	11%	5 yr		

> A. Which costs more in interest: a loan at 10% or 12%?
> B. Which costs less in interest: a $1000 loan at 8% for 2 years or a $900 loan at 7% for 3 years?

Commission

Sales people are often paid a commission on their total amount of sales.
The commission is usually expressed as a percent of sales.

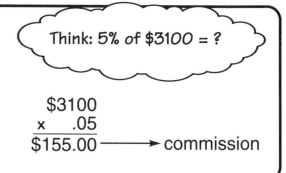

Julio earns a 5% commission on sales. Last week he sold $3100 in merchandise. How much commission should he receive?

Think: 5% of $3100 = ?

Commission = rate × sales
                5% × $3100

$3100
× .05
$155.00 ⟶ commission

1. Complete the chart to find the amount of commission.

Amount of Sales	Rate	Commission
$900.00	6%	_____
$350.00	12%	_____
$88.00	5%	_____
$749.35	7%	_____
$810.00	35%	_____
$120.00	3%	_____
$70,000	8%	_____

2. Lulu is in charge of recording sales commissions at the stereo shop. Below is her daily slip. Check her calculations and correct any errors she made.

Employee Code	Amount of Sales	Rate	Commission
A	$1900	5%	$95
B	$2430	5%	$125
C	$ 865	5%	$432.50
D	$3250	7%	$162.50
E	$2110	5%	$105.50

How would you evaluate Lulu's accuracy?

_____

3. A used car salesperson gets 6% commission on sales. What commission does the salesperson get on a car costing $5450?

_____

4. Real estate agents often get paid by commission. If the rate of commission is 7%, how much does the agent get paid for selling a home for $82,000?

_____

5. The Camera Club sold cards to raise money. They got 25% commission. How much did they raise on sales totaling $350?

_____

6. Jo's boss offered to pay her either a salary of $650 per month or a commission rate of 10%. She thinks her monthly sales would be about $7000. If this were correct, which method of payment would give Jo more money?

_____

Absolute Value of a Number

The absolute value of a number is its positive value. The symbol for absolute value is | |. The sign of the number (+ or −) is disregarded in absolute value.

The absolute value of +4 is 4.　　　The absolute value of −2 is 2.
　　　|+4| = 4　　　　　　　　　　　|−2| = 2
The absolute value of 0 is 0.
　　　|0| is 0.

1. A temperature rose 7˚. What is the absolute value of the change in temperature?

　　　_____

2. A temperature decreased 7˚. What is the absolute value of the change in temperature?

　　　_____

3. A stock increased in value 5 points. What is the absolute value of the change in stock price?

　　　_____

4. A stock decreased in value 6 points. What is the absolute value of the change in stock price?

　　　_____

Give the absolute value of each number.

5. |−7| _____　　6. |4| _____　　7. |0| _____　　8. |−5| _____

9. |3| = _____　　10. |−7| = _____　　11. |−3| = _____　　12. |0| = _____

Make a true statement by writing >, < or = in each circle.

13. |3| ◯ |−4|　　　14. |25| ◯ |150|　　　15. |0| ◯ |−5|

16. |−20| ◯ 0　　　17. |−13| ◯ |13|　　　18. |7| ◯ |−7|

Opposite and Absolute Value of a Number

Two different numbers, the same distance from zero on the number line, are opposites of each other.

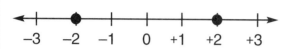

−2 is the opposite of +2
+2 is the opposite of −2

The absolute value of a number is the positive value of the number. The sign for absolute value is | |.

$$|-4| = 4 \quad |3| = 3 \quad |0| = 0$$

Name the opposites of each number.

1. −3 _____   **2.** +4 _____   **3.** −2 _____   **4.** 0 _____

5. $1\frac{1}{2}$ _____   **6.** −3.5 _____   **7.** $\frac{-5}{8}$ _____   **8.** 2.6 _____

Graph the number and its opposite on the number line.

9. 3
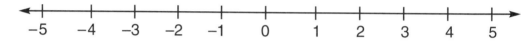

10. −4
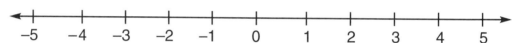

11. $2\frac{1}{2}$
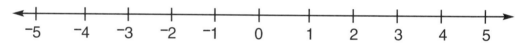

12. −3.5
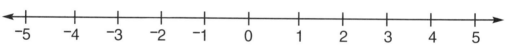

Find the absolute values.

13. |−5| _____   **14.** |7| _____   **15.** |−22| _____   **16.** |0| _____

17. |−50| _____   **18.** |27| _____   **19.** |−512| _____   **20.** |−100| _____

Which absolute value is greater?

21. |−6| or |−3| _____   **22.** |−12| or |+2| _____

23. |−100| or |80| _____   **24.** |−16| or |0| _____

Mathematical Reasoning

Table of Contents

Rounding to the Nearest 1000 Using a "Halfway" Number

You can use base ten blocks and a "halfway" number to help round numbers.
Here's how it works…

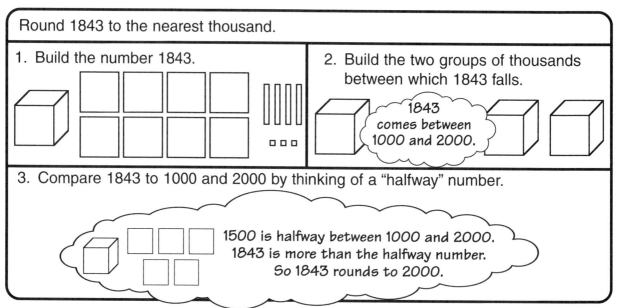

Round 1843 to the nearest thousand.

1. Build the number 1843.

2. Build the two groups of thousands between which 1843 falls.

1843 comes between 1000 and 2000.

3. Compare 1843 to 1000 and 2000 by thinking of a "halfway" number.

1500 is halfway between 1000 and 2000.
1843 is more than the halfway number.
So 1843 rounds to 2000.

Complete the table to help round the numbers to the nearest 1000.

	Number	Comes Between		Halfway Number	Rounds to
1.	1,287	1000 **and** 2000		1500	1000
2.	1,514	and			
3.	1,704	and			
4.	2,368	and			
5.	7,540	and			
6.	9,847	and			
7.	12,416	and			
8.	43,658	and			
9.	68,912	and			
10.	75,400	and			

11. Joni's mother earned $24,385 last year. How much money did she earn to the nearest thousand?

12. Jack earned $2,645 last year. How much did he earn to the nearest thousand?

13. The deepest part of the ocean is 36,198 feet. What is the depth to the nearest thousand?

14. There are 5,280 feet in a mile. To the nearest thousand, how many feet are in a mile?

Step 1 Read and understand.

Step 2 Find the question and needed facts.

Step 3 Decide on a process.

Step 4 Estimate.

Step 5 Solve and check back.

Read. Underline the question. Circle the needed facts. Complete the problem solving steps. Label the answer.

1. Toni is reading a book with 361 pages. So far she has read 178 pages. Jim is reading a book with 240 pages. How many more pages does Toni have to read?

   Process: + − × ÷

   Est. _____ Actual _____

2. The baseball team needs to collect 12,000 cans to raise money for new uniforms. Each uniform costs $50. They have collected 7,560 cans. How many more cans do they need?

   Process: + − × ÷

   Est. _____ Actual _____

3. The Golden Gate Bridge is 1,400 yards long. The George Washington Bridge is 3,500 feet long. Which bridge is longer? How much longer?

   Process: + − × ÷

   Est. _____ Actual _____

4. The stadium holds 12,480 people. On Tuesday 8,762 people came to the game. The hot dog vendors sold 1,560 hot dogs. How many empty seats were there?

   Process: + − × ÷

   Est. _____ Actual _____

5. A school has 425 girls, 413 boys and 34 teachers. How many students and teachers are in the school?

   Process: + − × ÷

   Est. _____ Actual _____

6. In a recent election 1,394 people voted for Paul, 942 voted for Paige and 847 voted for Ramona. How many more votes did Paige receive than Ramona?

   Process: + − × ÷

   Est. _____ Actual _____

7. Paul had 178 marbles. He gave away 23, then he gave away 47 more. How many marbles does Paul have?

   Process: + − × ÷

   Est. _____ Actual _____

8. The cost of material for a remodeling job was: lumber $1,240; hardware $312; paint $183. The labor charge was $3,200. What is the total cost for materials?

   Process: + − × ÷

   Est. _____ Actual _____

Problem Solving: Use Different Strategies to Decide on a Process

List the numbers of the strategies you can use to solve each problem. Estimate. Solve and check back.

1. If a car travels 58 miles per hour, how far will it go in 16 hours?

   Strategies _____

   Estimate _____ Actual _____

2. Krista and Heather completed a jogging race in 1080 seconds. How many minutes did they jog?

   Strategies _____

   Estimate _____ Actual _____

3. A skating rink sells an average of 706 tickets each day. How many tickets are sold in September and October?

   Strategies _____

   Estimate _____ Actual _____

4. A plane flew 3300 miles in 6 hours. How many miles per hour did it travel?

   Strategies _____

   Estimate _____ Actual _____

5. A car driven 140 miles used 5 gallons of gas. How many miles per gallon did the car average?

   Strategies _____

   Estimate _____ Actual _____

6. Alyse had 134 stamps. She put 8 stamps on each page. How many pages did she fill? How many stamps were left?

   Strategies _____

   Estimate _____ Actual _____

7. Jesse swims 12 laps each day. How many days will it take him to swim 280 laps?

   Strategies _____

   Estimate _____ Actual _____

8. Carrin needs 4 ft of material to make a table decoration. How many decorations can she make from 87 ft of material?

   Strategies _____

   Estimate _____ Actual _____

9. Fifty Scouts went camping. If one car holds 4 Scouts and their camping gear, how many cars will be needed?

   Strategies _____

   Estimate _____ Actual _____

10. The distance to the Scout camp is 130 miles. If a car averages 20 miles per gallon, how many gallons will be needed for one roundtrip?

    Strategies _____

    Estimate _____ Actual _____

Problem Solving: Read a Table, Estimate, Guess and Check

Nick's mother drives round-trip from Northville to Elmira three times a week. About how many miles does she drive?

Mileage Between Cities	Byron	Elmira	Northville	Sherman
Byron		128	318	451
Elmira	128		190	261
Northville	318	190		133
Sherman	451	261	133	

It's 190 miles from Northville to Elmira, about 200 miles. The round trip is about 400 miles.

$3 \times 400 = 1200$ miles per week estimating
$3 \times 380 = 1140$ miles actual

Estimating is using zeros with the basic facts... 3 x 4 plus 2 zeros.

Estimate and then find the actual number of miles in each trip.

1. Four one-way trips between Northville and Byron.

Estimate _____ Actual _____

2. One round trip between Byron and Sherman.

Estimate _____ Actual _____

3. A trip from Northville to Elmira and then from Elmira to Sherman.

Estimate _____ Actual _____

4. Two round trips between Sherman and Elmira.

Estimate _____ Actual _____

Guess and check with a calculator.

Calculation Game for Estimating Multiplication

49 62 21 39 31 51 8 12

3038	1302	612	744	312	372
96	248	1029	1989	651	1071
468	819	1519	2418	20	168
588	1922	496	2499	408	392
1911	4827	252	1581	3162	1209

1. Players take turns picking any two numbers from the cloud and multiplying those numbers on a calculator.

2. Cover the product on the game board with your marker (kernel of corn or bean).

3. First player with four-in-a-row is the winner.

Problem Solving: Find the Pattern to Estimate

1. Read.
2. Find question and needed facts.
3. Decide on process.
4. Estimate.
5. Solve and check back.

Use base ten blocks or a calculator to find the pattern for dividing multiples of ten.

Divide	Quotient	# of zeros in Quotient
600 ÷ 20		
900 ÷ 30		
4200 ÷ 70		
4500 ÷ 50		

Estimate the quotient:

$$39 \overline{)780}$$

780 rounds to 800.
39 rounds to 40.
800 ÷ 40 = ?

$$80\cancel{0} \div 4\cancel{0} = 20$$

Describe the pattern for dividing multiples of ten: _____

_____

Estimate the quotients in your head. You will not have time to work the problems.

1. $57 \overline{)432}$

2. $21 \overline{)834}$

3. $87 \overline{)271}$

4. $62 \overline{)478}$

5. $43 \overline{)359}$

6. $49 \overline{)347}$

7. $87 \overline{)2714}$

8. $39 \overline{)1723}$

9. $92 \overline{)1924}$

10. $70 \overline{)4328}$

11. $28 \overline{)6294}$

12. $63 \overline{)5412}$

13. 248 ÷ 49 = _____

14. 4612 ÷ 51 = _____

15. 2763 ÷ 38 = _____

Problem Solving: Make a Function Table, Find the Pattern

A rate compares two different kinds of quantities. The word "per" means for each unit. A function machine relates the two quantities by a rule to make a table.

Jake's mother walks 3 miles each hour or 3 miles per hour. How far does she walk in 2 hours? 3 hours? n hours?

The number of miles is always related at the same rate to the number of hours. The miles are a function of the hours.

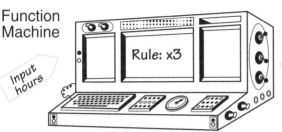

Function Machine

Rule: x3

Input hours

Output miles

Input Hours	Output Miles
1	3
2	$2 \cdot 3 = 6$
3	$3 \cdot 3 = 9$
n	$n \cdot 3 = 3n$

Find the rule for each problem. Use the rule to complete a table.

1. Jake rides his bike about 12 miles per hour. How far does he travel in 1 hour? 2 hours? 3 hours? n hours?

Rule: _____

Input (hour)	Output (miles)
1	
2	
3	
n	

2. Sound travels at a speed of 12 miles per minute. How far will it travel in 5 minutes? 10 minutes? 1 hour? n minutes?

Rule: _____

Input ( )	Output ( )

3. About 265 babies are born in the world every minute. How many babies are born in 2 minutes? 3 minutes? n minutes?

Rule: _____

Input ( )	Output ( )

4. Jake's brother waits tables at a local diner. He earns $6.50 an hour. How much does he earn in 4 hours? 8 hours? n hours?

Rule: _____

Input ( )	Output ( )

*Let the letter **d** represent the distance, **t** represent the time and **r** represent the rate. Use the letters **d**, **r** and **t** to write a number sentence that is always true. This number sentence is called the distance rule or formula.*

Justify Your Solution: Reasonable Answers with Remainders

Here are three different answers to the same problem.
Is one answer correct? Justify your decision.

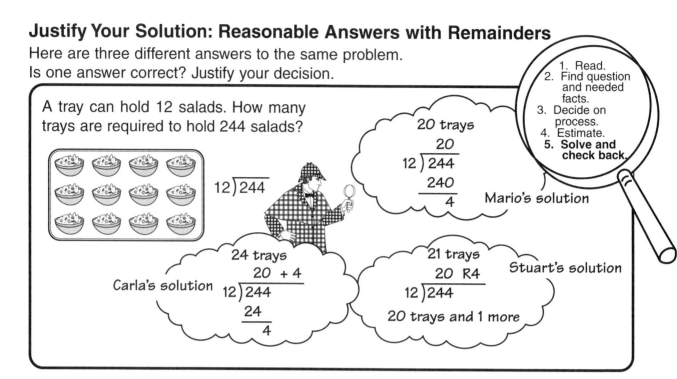

A tray can hold 12 salads. How many trays are required to hold 244 salads?

$12\overline{)244}$

20 trays
$$\begin{array}{r} 20 \\ 12\overline{)244} \\ 240 \\ \hline 4 \end{array}$$
Mario's solution

Carla's solution
24 trays
$$\begin{array}{r} 20\ +\ 4 \\ 12\overline{)244} \\ 24 \\ \hline 4 \end{array}$$

21 trays
$$\begin{array}{r} 20\ \ R4 \\ 12\overline{)244} \end{array}$$
20 trays and 1 more
Stuart's solution

1. Read.
2. Find question and needed facts.
3. Decide on process.
4. Estimate.
5. **Solve and check back.**

Solve. Write your answer in a complete sentence. Does it make sense?

1. A boat holds 5 people. How many boats are needed to hold 18 people?

2. 30 people are to be seated at tables of 4. How many tables will be needed?

3. Movie tickets cost $4. You have $25. How many tickets can you buy? How much will you have left?

4. The rows in a theater have 16 seats. How many rows are needed to seat 180 people?

5. A ferry takes 30 cars at a time to an island. How many trips will the ferry need to make if there are 143 cars waiting in line?

6. Hot dog buns are sold in packages of 10. You need 36 hot dogs for a cookout. How many packages must you buy? How many buns will be left over?

7. One dozen extra large eggs weigh 27 ounces. About how much does 1 egg weigh?

8. There are 42 children in a summer sports program. The children are divided into teams of 10. Any remaining students become substitutes. How many substitutes are there?

Problem Solving: Guess and Check, Draw a Picture, Make a Table

Joyce is 4 years older than Jesús. The sum of their ages is 22. Find their ages.

Joyce   Jesús

Joyce + Jesús = 22

Joyce has to be 4 years older than Jesús, and their ages must add up to 22. I'll make a table of guesses.

Joyce	Jesús	
5	1	5 + 1 = 6
10	6	10 + 6 = 16
12	8	12 + 8 = 20
13	9	13 + 9 = 22

Joyce is 13 years old.
Jesús is 9 years old.

Use guess and check to solve each problem.

1. A spaniel, Heidi, is 3 years older than a poodle named Muffin. The sum of their ages is 15. How old are the dogs?

   Heidi _____ Muffin _____

2. Mia is 4 years younger than her sister Shawn. The sum of their ages is 30. How old are the sisters?

   Mia _____ Shawn _____

3. You are twice as old as your brother. The sum of your ages is 18. What are your ages?

   You _____ Brother _____

4. Your father's age this year is a multiple of 5. Next year his age will be a multiple of 6. How old is your father now?

   _____

5. You have twice as many nickels as dimes. You have 80¢. How many nickels and dimes do you have?

   nickels _____ dimes _____

6. You have a collection of nickels and quarters worth $1.45. You have 1 more quarter than nickels. How many of each coin do you have?

   nickels _____ quarters _____

7. I am thinking of two numbers. If you add them, the sum is 34. Their difference is 6. What are the numbers?

   _____ _____

8. There are a total of 12 chickens and pigs in a barnyard. There are 38 animal feet in the yard. How many chickens and pigs are in the barnyard?

   chickens _____ pigs _____

Problem Solving: Using Logic to Eliminate Answers

A good problem solver uses clues about an unknown situation to solve the mystery. Careful reading and thinking allows you to use clues to eliminate solutions until you are left with the correct answer.

Guess My Sons' Ages
I have 2 sons. The product of their ages is my house number. The sum of their ages is less than 20 and is a multiple of the younger son's age. I live at 72 Elm St. How old are my sons?

Cross out all but the pairs whose sum is less than 20.

(1, 72) (2, 36) (3, 24) (4, 18) (6, 12) (8, 9)

Find the clue that eliminates the greatest number of possible answers...the product is 72. List the factors of 72.

Which pair has a sum that is the multiple of the younger son's age?
6 + 12 = 18 and 18 is a multiple of 6.
8 + 9 = 17 and 17 is not a multiple of 8.

(1, 72) (2, 36) (3, 24) (4, 18) (6, 12) (8, 9)

My sons' ages are 6 and 12.

1. Guess the number. It is a prime number. It is more than 20. The sum of its digits is 10. It is less than 50.

2. Guess the number. It is a three-digit whole number. Each of its digits is different. It is more than 800 and less than 900. It is an even number. It is divisible by 5. The sum of the digits is 12.

_____

_____

3. Steve, Jane, Raul and Irwin each play a different instrument: a clarinet, flute, saxophone or trumpet. Irwin plays the flute. Jane does not play the clarinet. Steve does not play the clarinet or saxophone. What does each person play?

4. Shane, Renee, Barbara and Carol each have different favorite sports: football, baseball, soccer and golf. Carol does not enjoy contact sports. Renee's favorite sport is played with a bat. Barbara does not like the injuries in football. What are their favorite sports?

Steve _____

Jane _____

Raul _____

Irwin _____

Shane _____

Renee _____

Barbara _____

Carol _____

Problem Solving: Work Backwards, Draw a Diagram

You have some money in your billfold. Your brother puts $20 more into the billfold and then you go shopping. You buy a gift for $18 and a card for $3. There is $14 left in the billfold. What was the original amount in the billfold?

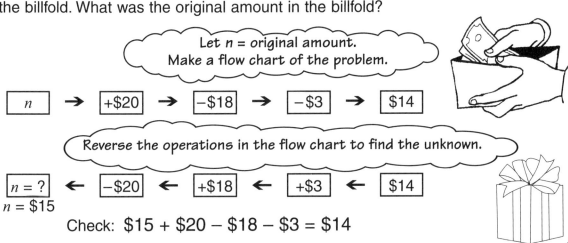

Let n = original amount.
Make a flow chart of the problem.

$$\boxed{n} \to \boxed{+\$20} \to \boxed{-\$18} \to \boxed{-\$3} \to \boxed{\$14}$$

Reverse the operations in the flow chart to find the unknown.

$$\boxed{n = ?} \leftarrow \boxed{-\$20} \leftarrow \boxed{+\$18} \leftarrow \boxed{+\$3} \leftarrow \boxed{\$14}$$
$$n = \$15$$

Check: $\$15 + \$20 - \$18 - \$3 = \$14$

Draw a flow chart to solve each problem. Check.

1. I am thinking of a number. If you multiply it by 6 and subtract 7, the answer is 35. What is the number?

_____

2. I am thinking of a number. If you add 30, subtract 10 and divide by 4, the answer is 20. What is the number?

_____

3. Diana had some pennies. She gave 10 pennies to her brother and then divided the leftover pennies equally between herself and 4 friends. Her share was 16¢. How many pennies did she start with?

_____

4. Dave bought some licorice for $1.80 a pound. He bought some trail mix for $4.00 and some bananas for $2.40. Dave spent a total of $10.00. How many pounds of licorice did he buy?

_____

There are 14 pennies in a pile. Two players take turns removing two or three pennies each turn. The player who takes the last penny is the winner. What is the best first move for the first player to make?

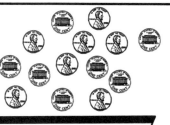

Number Sense, Mathematical Reasoning Check Point ☑

1. Write 250,000 in scientific notation.
(7NS 1.1)

2. What number goes in the box?
(7NS 1.2)

$$6 + (-8) = \boxed{}$$

3. Change $\dfrac{7}{10}$ to a percent.
(7NS 1.3)

4. A juice bar that used to sell for 50¢ now sells for 60¢. What is the percent of increase?
(7NS 1.6)

5. Tim bought a $500 personal computer. He received a 20% discount. What was the amount of discount?
(7NS 1.7)

6. Simplify. Write your answer in exponent form.
(7NS 2.1)
$$4^3 \times 4^2$$

7. Evaluate: $\dfrac{11}{12} - \dfrac{1}{3} =$
(7NS 2.2)

8. The value of $\sqrt{42}$ is between which two integers?
(7NS 2.4)

9. $|6| =$
(7NS 2.5)

10. Sue is taller than Sarah. Sarah is taller than Sam. What is a reasonable conclusion about the heights of Sue, Sarah and Sam?
(7MR 2.4)

A Sam is taller than Sue.

B Sarah is taller than Sue.

C Sue is taller than Sam.

D Sarah and Sue are the same height.

11. If the side of a square is doubled, what is the change in its area?
(7MR 3.3)

12. Which is not a reasonable estimate?
(7MR 2.1)

A $9,178 - 3,904 = 5,000$

B $794 - 386 = 400$

C $382 + 594 = 1000$

D $562 + 312 = 700$

Geometry and Measurement

Table of Contents

Inches, Feet, Yards

12 inches = 1 foot	3 feet = 1 yard
 (not actual size)	(not actual size)

Abbreviations: inch = in.   feet = ft.   yards = yd.

Complete.

1. 1 foot = _____ inches

2. 2 feet = _____ inches

3. 3 feet = _____ inches

4. 1 yard = _____ feet

5. 2 yards = _____ feet

6. 3 yards = _____ feet

7. 12 inches = _____ feet

8. 24 inches = _____ feet

9. 48 inches = _____ feet

10. 3 feet = _____ yards

11. 6 feet = _____ yards

12. 12 feet = _____ yards

13. 1 yard = _____ inches

14. 2 yards = _____ inches

15. 3 yards = _____ inches

16. 1 ft. 6 in. = _____ in.

17. 1 ft. 3 in. = _____ in.

18. 18 in. = _____ ft. _____ in.

19. 1 yd. 2 ft. = _____ ft.

20. 2 yd. 1 ft. = _____ ft.

21. 16 in. = _____ ft. _____ in.

Use a ruler and yardstick to find something in the room that is about an inch long, 1 foot, and 1 yard long.

Converting Length in the Metric System

Meter is the standard unit of length in the metric system.

Metric System Table of Length	Change 6 meters to centimeters.	Change 4500 m to km.
10 mm = 1 cm	*The answer should be greater, so multiply.*	*The answer should be less, so divide.*
100 cm = 1 m		$4500 \div 1000 = 4\underset{\smile}{5}\underset{\smile}{0}\underset{\smile}{0}$
1000 m = 1 km	$6 \times 100 = 600$ cm	= 4.5 km

What is the most sensible measure for the length of:

1. a new pencil

    18 mm   18 cm   18 m

2. a football

    25 mm   25 cm   25 m

3. a swimming pool

    15 cm   15 m   15 km

4. the distance between two cities

    18 cm   18 m   18 km

Complete.

                                    (multiply or divide)   (what number)

5. To change from cm to mm, _____ by _____

6. To change from m to cm, _____ by _____

7. To change from km to m, _____ by _____

8. To change from cm to m, _____ by _____

9. 3 cm = _____ mm

10. 40 mm = _____ cm

11. 65 mm = _____ cm

12. 4 km = _____ m

13. 3.5 km = _____ m

14. 2500 m = _____ km

15. 500 cm = _____ m

16. 9.5 m = _____ cm

17. 2.34 m = _____ cm

18. 4.2 cm = _____ mm

19. 83 mm = _____ cm

20. 4400 m = _____ km

Use a meter stick to measure:
A *The height of your desk* _____
B *The height of the classroom door* _____
C *The height of a wastebasket* _____

Customary Units of Weight

Weight means how heavy something is.
We measure weight in ounces (oz.) and pounds (lb.).

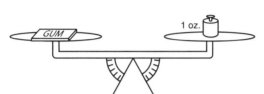 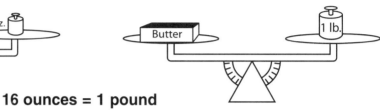

A pack of gum weighs about 1 ounce (oz.).

A package of butter weighs about 1 pound (lb.).

16 ounces = 1 pound
2,000 pounds = 1 ton

Would you measure in ounces, pounds or tons?

1. one quarter _____ **2.** a car _____ **3.** a person _____

4. a bushel _____ **5.** a watch _____ **6.** a bus _____
of apples

Rank the items from lightest (1) to heaviest (5).

7. an adult ____ paper clip ____ truck ____ book ____ desk ____

8. tennis ball ____ basketball ____ ping pong ball ____ football ____ baseball ____

Estimate the weight of:

9. an average bicycle **10.** one half gallon of milk **11.** an average school bus

_____ _____ _____

Complete.

12. 3 lb. = _____ oz. **13.** 80 oz. = _____ lb. **14.** $2\frac{1}{2}$ lb. = _____ oz.

15. 40 oz. = _____ lb. **16.** 3 ton = _____ lb. **17.** 8000 lb. = _____ ton

18. $2\frac{1}{2}$ ton = _____ lb. **19.** 6500 lb. = _____ ton **20.** 52 oz. = _____ lb.

21. 2.3 lb. = _____ oz. **22.** 7300 lb. = _____ ton **23.** 4.5 ton = _____ lb.

Units of Weight in the Metric System

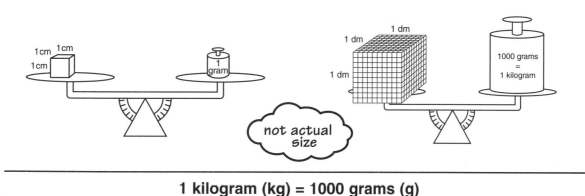

A one centimeter cube filled with water weighs about 1 gram (g).

A 10-centimeter cube filled with water weighs about 1000 grams or 1 kilogram (kg).

1cm 1cm 1cm

1 gram

not actual size

1 dm 1 dm 1 dm

1000 grams = 1 kilogram

1 kilogram (kg) = 1000 grams (g)

Would you measure in grams or kilograms?

1. a sugar cube

2. a dog

3. a golf ball

Which measurement is more reasonable?

4. a quarter: 5 g or 5 kg

5. an apple: 200 g or 200 kg

Complete.

6. 3 kg = _____ g

7. 2000 g = _____ kg

8. $2\frac{1}{2}$ kg = _____ g

9. 1500 g = _____ kg

10. 4.3 kg = _____ g

11. 1700 g = _____ kg

12. 1.25 kg = _____ g

13. 6500 g = _____ kg

14. 1250 g = _____ kg

15. Find the average of 5 kg, 7 kg, 13 kg and 11 kg.

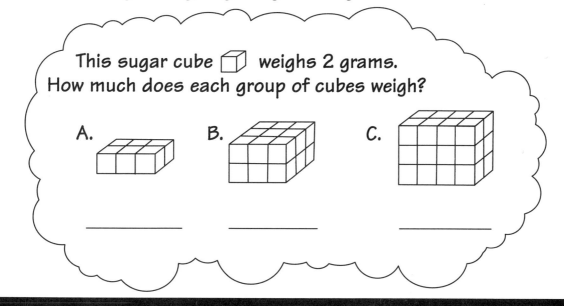

This sugar cube ⬜ weighs 2 grams.
How much does each group of cubes weigh?

A.

B.

C.

_____   _____   _____

Customary Units of Capacity

Capacity means how much something holds when filled up.
We often measure capacity in cups (c), pints (pt.), quarts (qt.) and gallons (gal.).

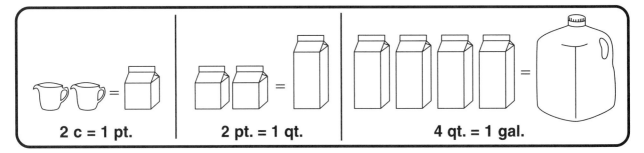

2 c = 1 pt.   **2 pt. = 1 qt.**   **4 qt. = 1 gal.**

Which is greater?

1. 2 c or 2 pt.   **2.** 2 qt. or 2 gal.   **3.** 4 qt. or 2 gal.

Complete.

4. 6 c = _____ pt.   **5.** 10 pt. = _____ qt.   **6.** 3 gal. = _____ qt.

7. 16 qt. = _____ gal.   **8.** 4 pt. = _____ c   **9.** 5 qt. = _____ pt.

10. 3 c = _____ pt.   **11.** 10 qt. = _____ gal.   **12.** $2\frac{1}{2}$ qt. = _____ pt.

13. A recipe calls for $\frac{1}{2}$ pt. sour cream. How many cups is this?

14. A punch recipe calls for 6 qt. of fruit juice. How many gallons is this?

_____

_____

15. One cup of milk makes one serving of hot chocolate. How many quarts of milk will be needed to make 28 servings?

16. A thermos container holds 2 gallons. How many cups does it hold?

_____

_____

Units of Capacity in the Metric System

We measure capacity in milliliters (mL) and liters (L). Small amounts of liquid are measured in milliliters. Larger amounts are measured in liters.

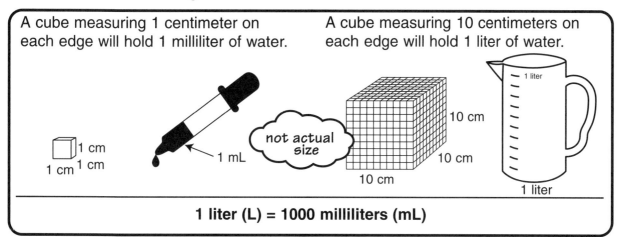

A cube measuring 1 centimeter on each edge will hold 1 milliliter of water.

A cube measuring 10 centimeters on each edge will hold 1 liter of water.

1 liter (L) = 1000 milliliters (mL)

Does each container hold more or less than a liter?

1.  _____

2. _____

3. _____

Would you measure how much each holds in milliliters or liters?

4. an eyedropper

5. a juice pitcher

6. the juice in a lemon

_____

_____

_____

Complete.

7. 2 L = _____ mL

8. 3000 mL = _____ L

9. 5 L = _____ mL

10. 6000 mL = _____ L

11. 4500 mL = _____ L

12. 2.5 L = _____ mL

13. 1 cm cube = _____ mL

14. 1 L = _____ cm cubes

15. 1 mL = ___ cm cube(s)

16. Tom bought 3 cans of orange juice. Each can held 250 mL. Did he buy more or less than one liter of orange juice?

_____

17. Mary bought 60.8 L of gas and 5.4 L of oil. How many liters did she buy in all?

_____

Measuring Temperature and Weight

We measure temperature with a thermometer marked in degrees (°).

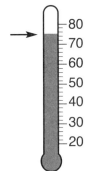

The temperature is 75°.

We measure weight with a scale.

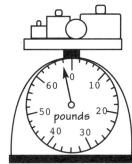

The scale reads 68 pounds.

What is the temperature?

1. _____

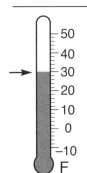

2. _____

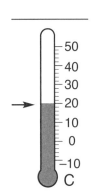

3. _____

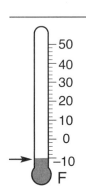

4. _____

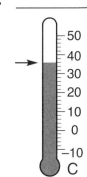

What is the weight?

5. _____

6. _____

7. _____

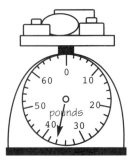

8. _____

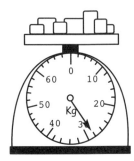

A Use a Fahrenheit thermometer to find the temperature of:
   (a) the room
   (b) a glass of water
   (c) a cup of hot water

B Use a Celsius thermometer to find the temperature of:
   (a) the room
   (b) a glass of water
   (c) a cup of hot water

Conversions in the Customary Measurement System

Length	Weight	Capacity	Time
12 in. = 1 ft.	16 oz. = 1 lb.	8 fluid oz. = 1 c.	60 sec. = 1 min.
3 ft. = 1 yd.	2000 lb. = 1 ton (T)	2 c. = 1 pt.	60 min. = 1 hr.
5280 ft. = 1 mi.		2 pt. = 1 qt.	24 hr. = 1 d.
		4 qt. = 1 gal.	7 d. = 1 wk.
			52 wk. = 1 yr. = 12 mo.

Use this table to solve these problems. You may wish to draw pictures to help you understand the relationships.

1. 2 lb. = _____ oz.

2. $6\frac{1}{2}$ ft. = _____ in.

3. 6 ft. = _____ yd.

4. 3 qt. = _____ pt.

5. 3 qt. = _____ c.

6. 1 d. = _____ min.

7. 6 wk. = _____ d.

8. 3 yr. = _____ mo.

9. 3 mi. = _____ ft.

10. 1 yd. = _____ in.

11. $2\frac{1}{2}$ ton = _____ lb.

12. 1 hr. = _____ sec.

13. 2 c. = _____ oz.

14. 5000 lb. = _____ ton

15. 16 c. = _____ pt.

16. 3 gal. = _____ qt.

17. 7 min. = _____ sec.

18. 12 hr. = _____ min.

19. 72 hr. = _____ d.

20. 64 oz. = _____ gal.

21. 1 mi. = _____ yd.

Give the abbreviation for each term.

22. minute _____

23. second _____

24. cup _____

25. week _____

26. quart _____

27. gallon _____

28. pint _____

29. pound _____

30. ton _____

31. foot _____

32. mile _____

33. yard _____

Some months have 31 days,

Some have 30 days.

How many have 28 days?

Converting Measurements in the Metric System

Length	Weight	Capacity
10 mm = 1 cm	1000 mg = 1 g	1000 mL = 1 L
10 cm = 1 dm	1000 g = 1 kg	1000 L = 1 kL
10 dm = 1 m		
1000 m = 1 km		

Use the tables to convert metric measurements. You may wish to draw pictures to show relationships to help you decide how to solve the problem.

1. 3 cm = _____ mm

2. 4 m = _____ cm

3. 5000 m = _____ km

4. 750 cm = _____ m

5. 3.4 km = _____ m

6. 7 dm = _____ cm

7. 3 L = _____ mL

8. 4000 mL = _____ L

9. 2.5 L = _____ mL

10. 5 g = _____ mg

11. 4.3 kg = _____ g

12. 6400 g = _____ kg

13. Mike is 1.85 meters tall and Kim is 1.92 meters tall. How many meters taller is Kim than Mike? _____ How many centimeters? _____

14. Dennis weighed 50.6 kilograms in August. In January he weighed 58.4 kilograms. How many kilograms had he gained? _____ How many grams? _____

15. A dictionary is 45 cm thick. How many centimeters thick are 10 dictionaries? _____ How many meters? _____

16. A large bottle of soda pop holds 1 liter. How many milliliters are there in 1 large bottle of soda pop? _____ How many milliliters in 4 large bottles of soda pop? _____

The Metric Prefix

The prefix (beginning part of a word) of a metric unit of measurement can help you convert from one unit of measurement to another. The most common prefixes are kilo, hecto, deka, deci, centi and milli, as in kilometer, hectometer, etc.

Use this chart to count the number of jumps from one prefix to another prefix. The abbreviation for each prefix is in parenthesis. Multiply the number by 10 each time you jump to the right. Divide the number by 10 each time you jump to the left.

kilo (k)   hecto (h)   deka (D)   gram (g)
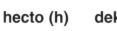
meter (m)   deci (d)   centi (c)   milli(m)
liter (L)

⟵ divide   multiply ⟹

Example 2500 milliliters = _____ liters	Example 3.6 kilometers = _____ meters
From mL to L is 3 jumps to the left. Each jump is a division by 10. 3 jumps is a division by 1000.	From km to m is 3 jumps to the right. Each jump is a multiplication by 10. 3 jumps is a multiplication by 10 x 10 x 10 = 1000.
$1000 \overline{)2500}$ = 2.5 or 2500 ÷ 1000 = 2.5	3.6 × 1000 = 3600 m

Write the word for each abbreviation.

1. cg _____
2. cm _____
3. dg _____

4. m _____
5. L _____
6. g _____

7. kg _____
8. mm _____
9. kL _____

Convert by jumping between prefixes.

10. 5 km = _____ m
11. 5 km = _____ dm
12. 5 km = _____ cm

13. 5 km = _____ mm
14. 6 g = _____ cg
15. 6 g = _____ mg

16. 7 kg = _____ g
17. 7 kg = _____ mg
18. 1.6 L = _____ dL

19. 1.6 L = _____ cL
20. 1.6 L = _____ mL
21. 8 dm = _____ cm

22. 18 kL = _____ L
23. 8 cm = _____ mm
24. 17 m = _____ cm

Miles Per Hour (m.p.h.)

Speed or how fast we travel is often given by telling how many miles are traveled in one hour. This is called "miles per hour" or m.p.h.

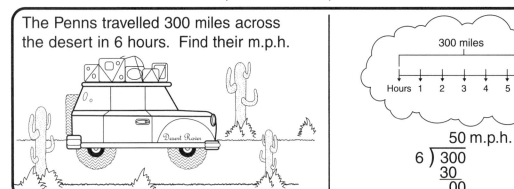

The Penns travelled 300 miles across the desert in 6 hours. Find their m.p.h.

300 miles

Hours 1 2 3 4 5 6

$$\begin{array}{r} 50 \text{ m.p.h.} \\ 6\overline{)300} \\ \underline{30} \\ 00 \end{array}$$

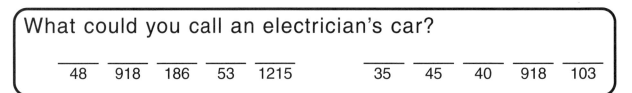

What could you call an electrician's car?

___ ___ ___ ___ ___   ___ ___ ___ ___ ___
48 918 186 53 1215   35 45 40 918 103

Find the m.p.h. Write the letter of the problem above the answer.

A. The Penns drove 270 miles up the California coast in 6 hours. Find the m.p.h.

_____

G. The Penns took 7 hours to drive 280 miles through small towns. Find the m.p.h.

_____

T. Martha drove 212 miles in 4 hours. Find her m.p.h.

_____

W. It took Tom 10 hours to drive 350 miles. Find the m.p.h.

_____

V. Jake drove 18 hours on a 3-day trip. He travelled 864 miles. Find the m.p.h.

_____

L. An airplane travels 744 miles in 4 hours. What is its speed?

_____

N. Richard Byrd flew 1545 miles on a 15-hour flight to the North Pole. Find his m.p.h.

_____

O. The Concorde jet flew the 3672 miles from London to Washington, D.C. in about 4 hours. Find its speed.

_____

U. The L.A./Anchorage speed record took 4 hours to travel 3394 miles. Find the m.p.h.

_____

S. The L.A./New York speed record took 3 hours to travel 3645 miles. Find the speed record.

_____

Unit Price: Finding the Better Buy

If the quality of the Brand A and Brand B peaches is equal, which is the better buy?

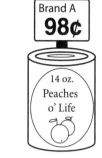

Brand A
98¢

14 oz.
Peaches
o' Life

Compute the unit price for each brand to find the better buy.

7¢ per oz.

14) 98

$8\frac{3}{4}$¢ per oz.

8) 70
    64
    6

Brand B
70¢

8 oz.
Peaches
& Dreams

Brand A is the better buy because it costs less per ounce.

Find the unit price. Round to the nearest cent when necessary.

1. Coffee: 32 oz. for $6.40

_____

2. Lemonade: 30 oz. for $3.29

_____

3. Peanut butter: 8 oz. for $1.75

_____

4. Popcorn: 16 oz. for $2.09

_____

Which is the better buy?

5. Flour: 5 lb. for $0.90
10 lb. for $1.70

6. Cheese: 12 oz. for $3.00
8 oz. for $2.00

7. Juice: 12 oz. for $0.86
32 oz. for $1.69

8. Crackers: 16 oz. for $1.19
9 oz. for $0.75

9. Cereal: 11 oz. for $1.87
16 oz. for $2.33

10. Raisins: 9 oz. for $0.98
24 oz. for $2.16

Solving Proportions with Models

Joan is making punch for a class party. She mixes 1 cup of lemon concentrate with every 3 cups of water. How many cups of water will be used with 4 cups of lemon concentrate?

Every cup of lemon concentrate will have 3 cups of water.

4 cups of lemon concentrate will have 12 cups of water.

Draw the missing figures to solve the proportion.

1. $\dfrac{\triangle}{\bigcirc\,\bigcirc\,\bigcirc\,\bigcirc} = \dfrac{\triangle\,\triangle}{\rule{2cm}{0.4pt}}$

2. $\dfrac{\triangle}{\bigcirc\,\bigcirc} = \dfrac{\rule{2cm}{0.4pt}}{\bigcirc\,\bigcirc\,\bigcirc\,\bigcirc\,\bigcirc\,\bigcirc}$

3. $\dfrac{\square\,\square\,\square}{\bigcirc\,\bigcirc\,\bigcirc\,\bigcirc} = \dfrac{\rule{2cm}{0.4pt}}{\bigcirc\,\bigcirc\,\bigcirc\,\bigcirc\,\bigcirc\,\bigcirc\,\bigcirc\,\bigcirc}$

4. $\dfrac{\triangle\,\triangle}{\rule{2cm}{0.4pt}} = \dfrac{\triangle}{\square\,\square\,\square}$

5. $\dfrac{\triangle\,\triangle}{|\,|\,|\,|\,|\,|\,|\,|\,|\,|\,|\,|} = \dfrac{\triangle}{\rule{2cm}{0.4pt}}$

6. $\dfrac{\bigcirc}{\triangle\,\triangle\,\triangle\,\triangle\,\triangle} = \dfrac{\bigcirc\,\bigcirc}{\rule{2cm}{0.4pt}}$

Draw figures to show each proportion. Solve for the missing number.

7. $\dfrac{1}{4} = \dfrac{\square}{8}$

8. $\dfrac{2}{3} = \dfrac{6}{\square}$

9. $\dfrac{2}{10} = \dfrac{1}{\square}$

10. $\dfrac{1}{6} = \dfrac{2}{\square}$

Scale Drawing

A map is a scale drawing. A scale drawing allows you to represent large distances on a small sheet of paper. The map of Treasure Island is drawn on a scale of 1 inch to 20 miles.

Use a ruler to answer the questions.

1. The distance between the East Gate and the West Gate on the map is 3 inches. What is the actual distance?

   _____

2. Find the actual distance from the treasure to these landmarks.

   a. Gopher Gulch   _____

   b. East Gate   _____

   c. West Gate   _____

   d. Lookout Point   _____

   e. Pirates Cove   _____

   f. Banyan Tree   _____

3. Find the actual distance from the Banyan Tree to Lookout Point.

   _____

4. Find the actual distance from the East Gate to Lookout Point.

   _____

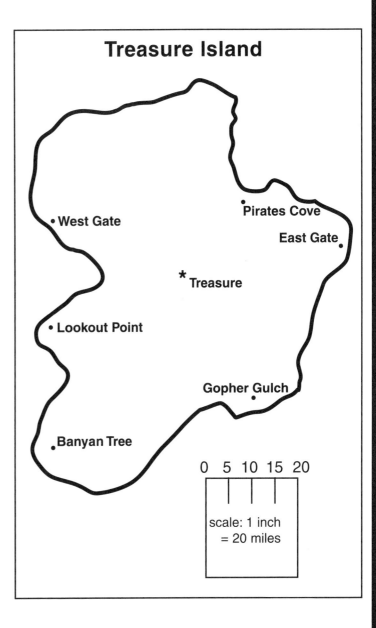

Treasure Island

- •West Gate
- •Pirates Cove
- East Gate•
- *Treasure
- •Lookout Point
- Gopher Gulch•
- •Banyan Tree

0 5 10 15 20

scale: 1 inch = 20 miles

Geometry Terms

Examples of ideas from geometry can be found everywhere in everyday life.

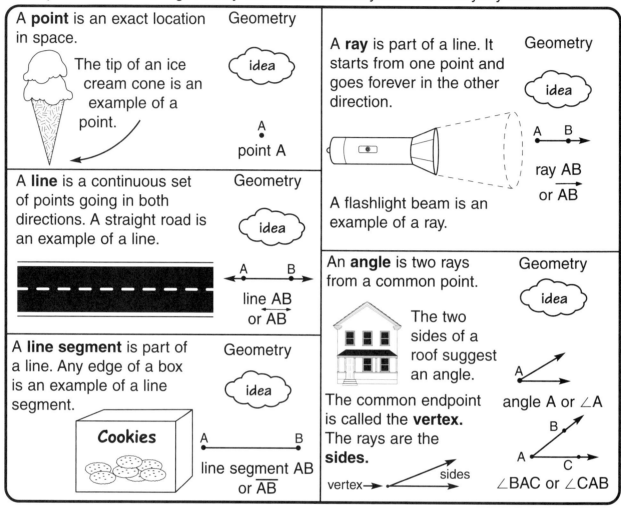

A **point** is an exact location in space.

The tip of an ice cream cone is an example of a point.

Geometry idea

A
•
point A

A **line** is a continuous set of points going in both directions. A straight road is an example of a line.

Geometry idea

A ← B →
line AB
or $\overleftrightarrow{AB}$

A **line segment** is part of a line. Any edge of a box is an example of a line segment.

Cookies

Geometry idea

A •———————• B
line segment AB
or $\overline{AB}$

A **ray** is part of a line. It starts from one point and goes forever in the other direction.

Geometry idea

A flashlight beam is an example of a ray.

A B →
ray AB
or $\overrightarrow{AB}$

An **angle** is two rays from a common point.

Geometry idea

The two sides of a roof suggest an angle.

The common endpoint is called the **vertex.** The rays are the **sides.**

vertex → • sides

A •————→
angle A or ∠A

B
A •————• C
∠BAC or ∠CAB

Point, line or line segment?

1. a window frame

2. where two streets meet

3. a view of an ocean

4. edge of a book

Ray, angle or neither?

5. beam of a flashlight

6. an open pair of scissors

7. cooked spaghetti

8. view from a telescope

Parts of a Circle: Center, Radius, Diameter, and Circumference

A bicycle wheel is an example of a circle and parts of a circle.

Everyday life Geometry	A **circle** is a set of continuous points equally distant from a point called the **center.** The distance around the outside of a circle is called the **circumference.**
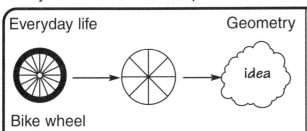 Bike wheel	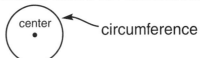
A **radius** is any line segment from the center to the circumference. $\overline{OM}$ is a radius of the circle.	A **diameter** is a line segment passing through the center with two endpoints on the circumference. 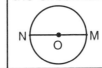 $\overline{NOM}$ is a diameter of the circle.

Center, radius, diameter or circumference?

1. rim of a basketball hoop _____

2. spoke of a wheel _____

3. scored line on an aspirin _____

4. path of a propeller _____

5. minute hand of a clock _____

6. rim of a circular wastebasket _____

What part of each circle is represented by the dotted line?

7. _____

8. _____

9. _____

10. Describe how a circle and a line segment are alike and how they are different.

_____

_____

11. Describe how a radius and a diameter are alike and how they are different.

_____

_____

12. If $\overline{QO}$ = 4 cm, then $\overline{QR}$ = _____.

13. If $\overline{QO}$ = 4 cm, then $\overline{OS}$ = _____.

14. Name 3 radii _____ _____ _____.

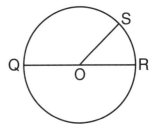

I am the longest straight line of a circle. What am I?

Kinds of Lines:
Horizontal, Vertical, Diagonal, Parallel, Intersecting, Perpendicular

diagonal – lines running at a slant

horizontal – lines running across

vertical – lines running up and down

intersecting lines – meet at a point
parallel lines – never intersect
perpendicular lines – intersect at right angles

Diagonal Blvd. intersects J Street and K Street.
J Street and K Street are parallel.
Third Ave. is perpendicular to J St. and K St.

1. How does the distance between J and K Street at 3rd Ave. compare to the distance between J and K Street at 4th Ave.?

   _____

   _____

2. Describe how J Street and Diagonal Blvd. are related.

   _____

   _____

3. Describe how K Street and 3rd Avenue are related.

   _____

   _____

4. Describe how you think Diagonal Blvd. and 3rd Avenue are related. Why?

   _____

   _____

Describe an example of each kind of line in this room.

5. horizontal

   _____

6. vertical

   _____

7. diagonal

   _____

8. parallel lines

   _____

9. intersecting lines

   _____

10. perpendicular lines

   _____

11. Draw $\overleftrightarrow{AB}$ parallel to $\overleftrightarrow{CD}$ in horizontal position.

12. Draw $\overleftrightarrow{MN}$ in a vertical position and a slanted line $\overleftrightarrow{OP}$ intersecting it.

13. Draw $\overleftrightarrow{JK}$ in a diagonal position and $\overleftrightarrow{MN}$ perpendicular to it.

14. How would you describe two parallel lines to a friend on the telephone?

   _____

15. How would you describe two perpendicular lines to a friend on the telephone?

   _____

Measuring Angles

The size of an angle is measured in units called **degrees.** You can think of one degree (1°) as being a very small wedge.

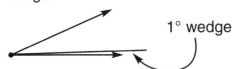

1° wedge

The size of an angle is the number of degrees (or wedges) between the two sides of the angle.

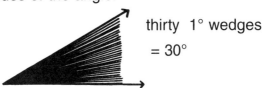

thirty 1° wedges
= 30°

A protractor is an instrument used to measure and draw angles. The protractor is divided into 180°.

The angle measured in this protractor shows 75° and 105°. Which is correct? How do you know?

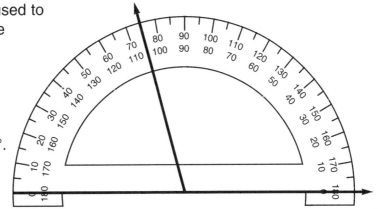

Estimate the number of degrees in each angle. Use a protractor to measure each angle. Compare your estimate to the actual measurements.

1. Estimate _____ Actual _____   **2.** Estimate _____ Actual _____

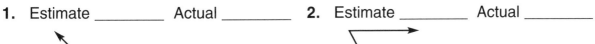

3. Estimate _____ Actual _____   **4.** Estimate _____ Actual _____

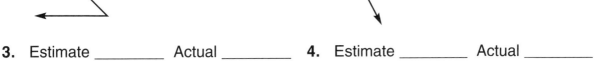

5. Estimate _____ Actual _____   **6.** Estimate _____ Actual _____

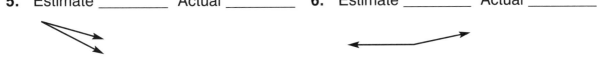

Estimate the number of degrees between the two hands of a clock when the time shown is:

7. 4 o'clock _____   **8.** 1 o'clock _____   **9.** 3 o'clock _____   **10.** 6 o'clock _____

Kinds of Angles: Right, Acute, Obtuse, and Straight

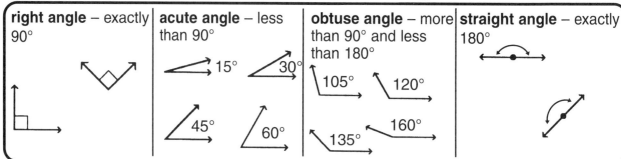

| right angle – exactly 90° | acute angle – less than 90° | obtuse angle – more than 90° and less than 180° | straight angle – exactly 180° |

Compare each angle to 90°. >, <, =

1. _____

2. _____

3. _____

4. _____

Describe each figure.

5.
_____
_____
_____

6. _____
_____
_____

7.
_____
_____
_____

8. _____
_____
_____

Acute, right, obtuse or straight?

9. _____

10. _____

11. _____

12. _____

13.
_____

14.
_____

15.
_____

16.
_____

What kind of angle is formed by the hands of a clock when the time is:

17. 3:00 _____

18. 5:00 _____

19. 6:00 _____

20. 10:00 _____

Every hour the minute hand of a clock moves and forms a "round" angle. How many degrees in a round angle?

Equilateral, Isosceles and Scalene Triangles

A figure formed by joining three points (not in a straight line) is a **triangle**. 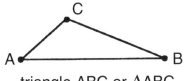 triangle ABC or △ABC	**Equilateral triangle** – all sides are equal. 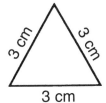
Isosceles triangle – only two sides are equal.	**Scalene triangle** – no sides are equal. 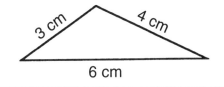

Name each triangle as equilateral, isosceles or scalene.

1.

_____

2.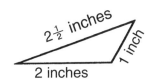

_____

3.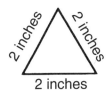

_____

4.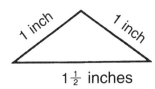

_____

5.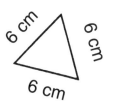

_____

6.

_____

7.

_____

8.

_____

9.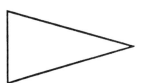

_____

How many angles are equal in an equilateral triangle? An isosceles triangle?   A scalene triangle?

Naming a Triangle by its Angles – Right, Obtuse, Acute

right triangle – a triangle that has a right angle	**obtuse triangle** – a triangle that has an obtuse angle	**acute triangle** – a triangle that has only acute angles
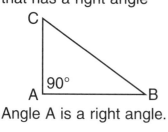	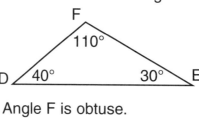	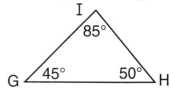
Angle A is a right angle.	Angle F is obtuse.	Angle G, H, I are acute.
△ ABC is a right triangle.	△ DEF is an obtuse triangle.	△ GHI is an acute triangle.

Name each triangle as right, obtuse or acute.

1.
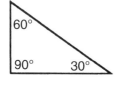

_____

2.
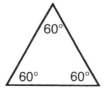

_____

3.
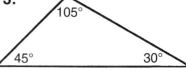

_____

4.
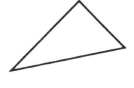

_____

5.
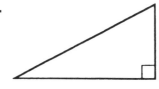

_____

6.
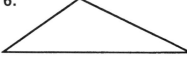

_____

7. Draw obtuse △ XYZ.

8. Draw right △ MNO.

9. Draw acute △ DEF.

10. Can you draw a right triangle that has an obtuse angle?

_____

11. Can you draw a triangle with two obtuse angles?

_____

AB and BC are equal because they are opposite equal angles. Is △ ABC equilateral, isosceles or scalene?

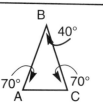

Naming Polygons

A figure made by connecting 3 or more points (not in a straight line) is a **polygon.**
Polygons are named by the number of sides.

triangle	quadrilateral	pentagon	hexagon
3 points – 3 sides	4 points – 4 sides	5 points – 5 sides	6 points – 6 sides
octagon	**decagon**	**dodecagon**	**regular polygon –**
8 points – 8 sides	10 points – 10 sides	12 points – 12 sides	all sides and angles are congruent

1. All snowflakes are hexagons. What common street sign is an octagon?

_____

2. What kind of polygon will most kites be?

_____

3. Construct a pentagon.

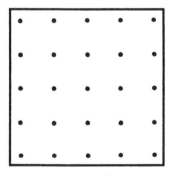

4. Construct an octagon.

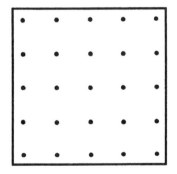

5. Construct a quadrilateral.

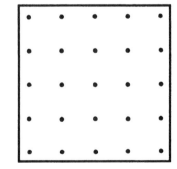

6. Construct a decagon.

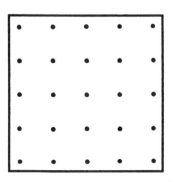

7. Construct a hexagon.

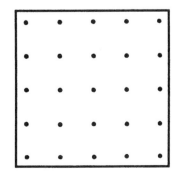

8. Construct a regular quadrilateral.

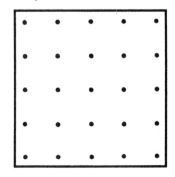

Special Quadrilaterals and Their Properties

These special **quadrilaterals** have properties that identify them.

parallelogram	rectangle	square	trapezoid

Describe each figure by using as *many* properties as possible. Write the letter of each property in the blank.

1. square _____

2. parallelogram _____

3. quadrilateral _____

4. trapezoid _____

5. rectangle _____

Describe each figure using as *few* properties as necessary.

6. square _____

7. parallelogram _____

8. quadrilateral _____

9. trapezoid _____

10. rectangle _____

PROPERTY BOX

A. There are 4 sides.

B. Opposite sides are equal.

C. Two pair of parallel sides.

D. Only one pair of parallel sides.

E. All angles are right angles.

F. All sides are equal.

G. Opposite angles are equal.

H. At least one angle is not a right angle.

I. One side is longer than another side.

J. At least one side is parallel to its opposite side.

K. There are 4 angles.

11. Describe a rectangle to a friend. _____
_____

12. Describe a parallelogram to a friend. _____
_____

13. Describe a square to a friend. _____
_____

14. Describe a trapezoid to a friend. _____
_____

Congruent Lines and Angles

Congruent figures have the same shape and size. Congruent figures look exactly alike but they don't have to be in the same position.

Congruent line segments have the same length.

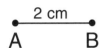 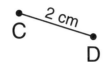

$\overline{AB}$ is congruent to ($\cong$) $\overline{CD}$
$\overline{AB} \cong \overline{CD}$

Congruent angles have the same measure.

$\angle A \cong \angle B$

Note: The size of an angle is not affected by the length of its sides.

---

1. Which line segments look congruent? _____

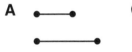

2. Which angles look congruent? _____

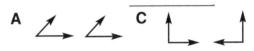

3. Which letters are congruent? _____

4. Which letters are congruent? _____

Use your ruler to draw a line congruent to the given line.

5.

6.

Use your protractor to draw an angle congruent to the given angle. You may have to extend the sides of the angle to measure it.

7.

8.

How many words can you make from the letters in
c-o-n-g-r-u-e-n-t?

Congruent Polygons from Slides, Turns or Flips

Congruent figures have the same shape and size but need not be in the same position.

Use a piece of tracing paper to copy this figure. Move the paper in different ways to show how the figure looks after a slide, turn or flip.

 SLIDE   TURN   FLIP

      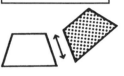

Trace each figure. Move the tracing paper to show the figure in a new position. Draw its new position.

1.

2.

3.

4.

5.
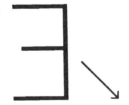

6.
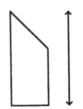

Corresponding Parts of Congruent Figures

Congruent triangles ABC and DEF have matching parts that fit exactly over each other. We call these matching parts **corresponding parts**.

△ABC ≅ △DEF

The matching sides are called **corresponding sides.** The matching angles are called **corresponding angles.**

∠A corresponds to (⟷) ∠D

$\overline{AB}$ ⟷ $\overline{DF}$

Which angle corresponds to ∠B? ∠E?

Which side corresponds to $\overline{BC}$? $\overline{AC}$?

The two figures are congruent. Give the corresponding parts.

1.

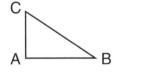

∠A ⟷ ___ $\overline{AB}$ ⟷ _____

∠B ⟷ ___ $\overline{BC}$ ⟷ _____

∠C ⟷ ___ $\overline{AC}$ ⟷ _____

If ∠B = 50°, ∠F = _____

2.

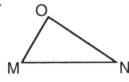

∠P ⟷ ___ $\overline{PQ}$ ⟷ _____

∠Q ⟷ ___ $\overline{PR}$ ⟷ _____

∠R ⟷ ___ $\overline{QR}$ ⟷ _____

If $\overline{MN}$ = 3 cm, PQ = _____

3.

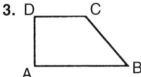

∠A ⟷ ___ $\overline{AB}$ ⟷ _____

∠B ⟷ ___ $\overline{BC}$ ⟷ _____

∠C ⟷ ___ $\overline{CD}$ ⟷ _____

∠D ⟷ ___ $\overline{AD}$ ⟷ _____

If ∠B = 70°, ∠H = _____

4.

∠N ⟷ ___ $\overline{MN}$ ⟷ _____

∠O ⟷ ___ $\overline{NO}$ ⟷ _____

∠M ⟷ ___ $\overline{MO}$ ⟷ _____

If $\overline{KL}$ = 15 in., $\overline{MO}$ = _____

5. Describe the corresponding parts of two congruent triangles to a friend.

_____

_____

_____

Coordinate Grid

This graph is made of two number lines of the set of integers (...−5,−4, −3, −2, −1, 0, 1, 2, 3, 4, 5...). The horizontal number line is called the **x axis** and the vertical number line is called the **y axis.** The point where x and y intersect is called the **origin** and is identified by the point (0,0).

The x and y axes separate the coordinate plane into four quadrants labeled I, II, III and IV. When identifying ordered pairs, first count how far right or left of zero and then how far up or down from zero.

Point A is located 3 spaces right or east and two spaces down or south. Point A is represented by the ordered pair (3, −2). What ordered pair identifies point B?

_____

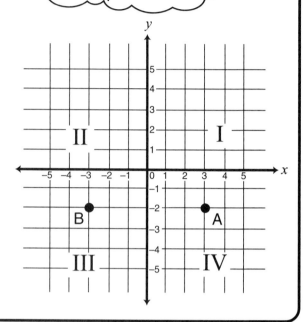

Remember the first number goes right or left.

The second number goes up and down.

Graph and label the points in the grid.

1.

Letter	Integer Pair
A	(4, −1)
B	(−2, 3)
C	(−3, 4)
D	(0, 3)
E	(−4, −3)
F	(1, 1)
G	(−2, −3)
H	(−1, −2)

2.

Letter	Integer Pair
P	(4, −2)
Q	(−3, 1)
R	(2, 0)
S	(0, −3)
T	(−4, −2)
U	(−5, 0)
V	(3, 2)
W	(0, 5)

3. In which quadrant do the following points lie?

(3, −2)   (−3, 2)   (3, 2)   (−3, −2)

_____ _____ _____ _____

4. What figure is formed by lines connecting the points in problem 3?

_____

Translations and Reflections

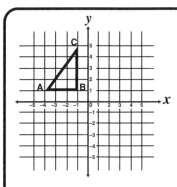

Triangle ABC is a right triangle.
Name the coordinates of each point:

A (___ , ___) B (___ , ___) C (___ , ___)

translation of a triangle

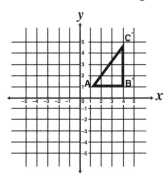

reflection of a triangle

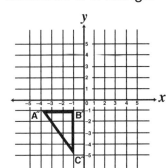

Coordinates

A′ (___ , ___)

B′ (___ , ___)

C′ (___ , ___)

Coordinates

A″ (___ , ___)

B″ (___ , ___)

C″ (___ , ___)

1. Name triangle DEF according to its sides.
Draw a triangle D′E′F′ that is a translation of triangle DEF.
Give the coordinates of each new point.

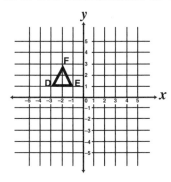

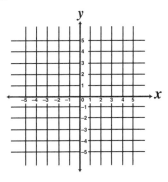

Coordinates

D′ (___ , ___)

E′ (___ , ___)

F′ (___ , ___)

2. Name triangle PQR according to its angles.
Draw a triangle P′Q′R′ that is a reflection of triangle PQR.
Give the coordinates of each new point.

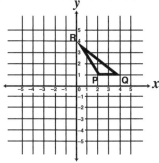

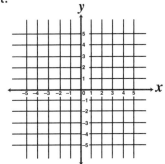

Coordinates

P′ (___ , ___)

Q′ (___ , ___)

R′ (___ , ___)

The Pythagorean Theorem

Pythagoras, a Greek mathematician, discovered a special property about right triangles. This property relates to the square that can be drawn on each side.

The right triangle below has sides of 3, 4 and 5.

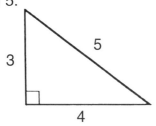

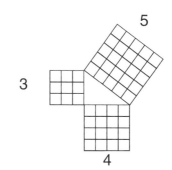

The shorter sides, 3 and 4, are called the **legs** of the right triangle. The longest side, 5, is called the **hypotenuse.** The hypotenuse is the side opposite the right angle.

$3^2 =$ _____ $4^2 =$ _____ $5^2 =$ _____

$3^2 + 4^2 =$ _____ $5^2 =$ _____

Describe this relationship (known as Pythagorean Theorem):

_____

_____

Three sides of a triangle are given. Is the triangle a right triangle?

1. 5, 12, 13   **2.** 4, 5, 6   **3.** 6, 8, 10

4. 5, 7, 9   **5.** 9, 12, 15   **6.** 7, 24, 25

Find the legs and hypotenuse of the right triangle formed by these squares.

7. legs = _____ hypotenuse = _____   **8.** legs = _____ hypotenuse = _____

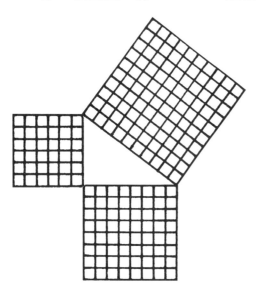

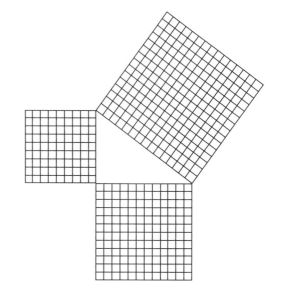

The Pythagorean Theorem to Find a Missing Side

The **Pythagorean Theorem:** The sum of the squares of the legs is equal to the square of the hypotenuse.

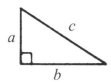

Leg a squared + leg b squared equals hypotenuse c squared.

We write: $a^2 + b^2 = c^2$

You can use the Pythagorean Theorem to find the missing side of any right triangle.

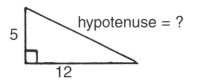

$a^2 + b^2 = c^2$ If $c^2 = 169$,

$5^2 + 12^2 = c^2$ $c = \sqrt{169}$

$25 + 144 = c^2$ $c = 13$

$169 = c^2$

1. $c^2 = 25$, $c =$____   2. $c^2 = 100$, $c =$____   3. $c^2 = 625$, $c =$____   4. $c^2 = 225$, $c =$____

Find the hypotenuse.

5.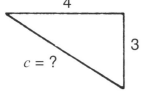

$a^2 + b^2 = c^2$

____ + ____ = ____

$c^2 =$ ____

$c =$ ____

6.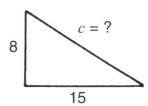

$a^2 + b^2 = c^2$

____ + ____ = ____

$c^2 =$ ____

$c =$ ____

7.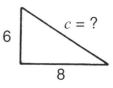

$a^2 + b^2 = c^2$

____ + ____ = ____

$c^2 =$ ____

$c =$ ____

Find the length of the missing leg.

8.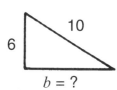

$a^2 + b^2 = c^2$

____ + ____ = ____

$b^2 =$ ____

$b =$ ____

9.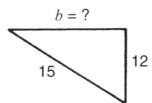

$a^2 + b^2 = c^2$

____ + ____ = ____

$b^2 =$ ____

$b =$ ____

10.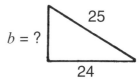

$a^2 + b^2 = c^2$

____ + ____ = ____

$b^2 =$ ____

$b =$ ____

Finding the Perimeter of a Polygon

Perimeter means the total distance around the outside.

Jack walked all the way around a football field. How many feet did Jack walk?

300 ft.

Jack starts here.

160 ft.

Jack walks 300 ft. and turns. He walks 160 ft. and turns. He walks 300 ft. and turns. He walks 160 ft. and is back where he started.

300 + 160 + 300 + 160 = 920
The perimeter of the football field is 920 ft.

The space between two dots represents 1 foot. Find the perimeter of each polygon.

1.

perimeter = _____ ft.

2.

perimeter = _____ ft.

3.

perimeter = _____ ft.

4.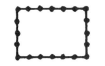

perimeter = _____ ft.

5.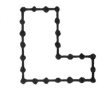

perimeter = _____ ft.

6.

perimeter = _____ ft.

The measurements of each side of the polygons are given. Find the perimeter.

7.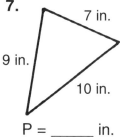

7 in.

9 in.

10 in.

P = _____ in.

8.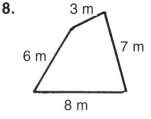

3 m

6 m   7 m

8 m

P = _____ m

9.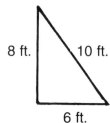

8 ft.   10 ft.

6 ft.

P = _____ ft.

10.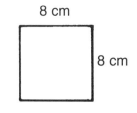

8 cm

8 cm

P = _____ cm

11.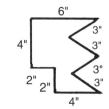

6"

4"   3"
        3"

2" 2"   3"
        3"

4"

P = _____ in.

12.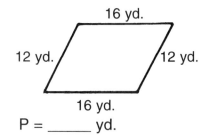

16 yd.

12 yd.   12 yd.

16 yd.

P = _____ yd.

Perimeter of a Triangle or Square with a Formula

Find the perimeter of a triangle with sides of 2.8 cm, 3 cm and 4.2 cm.	Find the perimeter of a square with sides of 3 cm.
1. Sketch	1. Sketch
$a = 2.8$ $b = 3$ $c = 4.2$	$s = 3$ cm (all four sides)
2. Formula $P = a + b + c$	2. Formula $P = s + s + s + s$ or $P = 4s$
3. Substitute $P = 2.8 + 3.0 + 4.2$	3. Substitute $P = 4 \times 3$
4. Solve $P = 10.0$ cm	4. Solve $P = 12$ cm

Use the four steps to find the perimeter of a triangle with the given dimensions.

1. sides of 56 ft., 17 ft., and 67 ft.

2. sides of 0.53 m, 0.46 m and 0.85 m

3. sides of $2\frac{1}{4}$ in., 3 in., and $4\frac{1}{2}$ in.

Use the four steps to find the perimeter of a square with the given dimensions.

4. sides of 25 ft.

5. sides of 11 km

6. sides of 6.5 in.

7. sides of $3\frac{3}{4}$ in.

8. sides of 3.68 km

9. sides of $\frac{1}{3}$ ft.

Using a Formula to Find the Perimeter of a Rectangle

Follow four steps to find the perimeter of a rectangle with length of 7 inches and width of 4 inches.

1. Draw a sketch.

$l = 7$ in.

$w = 4$ in. [rectangle] $w = 4$ in.

$l = 7$ in.

$P = (2 \times 7) + (2 \times 4)$
2. Find a formula from the sketch.
   $P = (2 \text{ lengths}) + (2 \text{ widths})$
   $P = (2l) + (2w)$

3. Substitute values.
   $P = (2l) + (2w)$

   Take out *l* and put in 7.
   Take out *w* and put in 4.

4. Solve and label.
   $P = (2l) + (2w)$
   $= (2 \times 7) + (2 \times 4)$
   $= 14 + 8$
   $= 22$ inches

Follow four steps to find the perimeter of a rectangle with the given dimensions.

1. length of 125 km, width of 90 km

2. length of 3.06 miles, width of 2 miles

3. length of .4 m, width of .3 m

4. length of $2\frac{1}{2}$ in., width of $1\frac{1}{4}$ in.

5. length of $3\frac{1}{5}$ km, width of $2\frac{3}{10}$ km

6. length of 8 yd., width of 2.5 yd.

Circumference of a Circle

The distance around the outside of a circle is called its **circumference.**

1. Use a piece of string and a centimeter ruler to find the circumference of this circle to the nearest centimeter.
    circumference = _____ cm

2. About how many times larger is the circumference than the diameter?

    _____

3. If the diameter of a circle is 14 cm, guess what its circumference might be _____

4. The circumference of a circle is always equal to its diameter multiplied by $3\frac{1}{7}$ or 3.14. This constant value is called pi (π).
    Can you write a formula for the circumference of a circle (use C, π, d and $=$)?
    Formula: _____

d = 7 cm

Find the circumference of each circle. Use $3\frac{1}{7}$ for π.

5.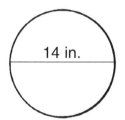

6.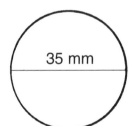

7.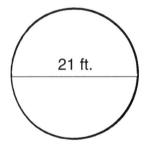

$C =$ _____ in. $C =$ _____ mm $C =$ _____ ft.

Find the circumference of each circle. Use 3.14 for π.

8.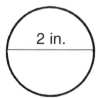

9.

10.

$C =$ _____ in. $C =$ _____ cm $C =$ _____ m

Using a Formula to Find the Circumference of a Circle

Find the circumference of a circle with diameter of 4.5 cm.

1. Sketch

4.5 cm

2. Formula $C = \pi d$
3. Substitute $= 3.14 \times 4.5$ cm

> Use 3.14 for π because the diameter is a decimal.

4. Solve and label $= 14.13$ cm

Find the circumference of a circle with diameter of $3\frac{1}{2}$ in.

1. Sketch

$3\frac{1}{2}$ in.

2. Formula $C = \pi d$
3. Substitute $= 3\frac{1}{7} \times 3\frac{1}{2}$ in.

> Use $3\frac{1}{7}$ for π because the diameter is a fraction.

4. Solve and label $= \frac{22}{7} \times \frac{7}{2}$
                     $= 11$ in.

Use the four steps to find the circumference of a circle with given diameter. Use $3\frac{1}{7}$ or 3.14 for π.

1. $d = 2$ in.

$C =$ _____ in.

2. $d = 18$ m

$C =$ _____ m

3. $d = 1.1$ ft.

$C =$ _____ ft.

4. $d = 2.5$ mm

$C =$ _____ mm

5. $d = 5.2$ dm

$C =$ _____ dm

6. $d = 4\frac{1}{2}$ in.

$C =$ _____ in.

7. $d = 14$ miles

$C =$ _____ miles

8. $d = 4\frac{2}{3}$ yd

$C =$ _____ yd.

9. $d = 5\frac{1}{4}$ ft.

$C =$ _____ ft.

A Formula for the Area of a Rectangle

Juanita is tiling a bathroom floor with 1-foot squares of ceramic. The floor is 8 feet long and 6 feet wide. How many squares will she need?

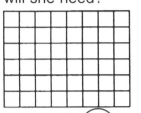

There will be 6 rows...
Each row will have 8 one-foot squares...
6 x 8 or 48 squares will be used.

A **formula** is a pattern or rule that is always true. You can find the area of a rectangle by counting tiles in the picture or multiplying length times width.

Area = length times width

Formula: $A = lw$

$A = 8 \times 6$

$= 48$ square ft.

Use the four steps (sketch, formula, substitute and solve) to find the area of each rectangle. Be sure to label your answer with square units.

1. length = 15 in.
width = 4 in.

2. length = 5 cm
width = 3 cm

3. length = 12 ft.
width = 10 ft.

$A =$ _____ square in.

$A =$ _____ square cm

$A =$ _____ square ft.

4. $l = 3.4$ miles
$w = 2$ miles

5. $l = 7.8$ ft. $w = 2.1$ ft.

6. $l = 5$ m $w = 2.3$ m

$A =$ _____

$A =$ _____

$A =$ _____

7. $l = \frac{1}{2}$ in. $w = \frac{1}{4}$ in.

8. $l = 8$ ft. $w = \frac{1}{2}$ ft.

9. $l = 4\frac{1}{2}$ in. $w = 2$ in.

$A =$ _____

$A =$ _____

$A =$ _____

A Formula for the Area of a Square

A square is a special kind of rectangle in which all sides are equal. Instead of multiplying length times width to find the area, we think of multiplying side times side.

Find a formula for the area of a square. Use the formula to find the area of a square with 5 cm on each side.

There are 5 rows . . . . There are 5 small squares in each row . . . .

side = 5 cm

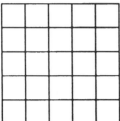

side = 5 cm

Area = side times side

Formula: $A = s \times s$ or $A = s^2$

Solve: $A = 5^2$

$= 5 \times 5$

$= 25$ square cm

Use the four steps (sketch, formula, substitute and solve) to find the area of each square.

1. side = 3 in.

$A = $ _____ _____

2. side = 6 m

$A = $ _____ _____

3. side = 12 in.

$A = $ _____ _____

4. side = 0.5 km

$A = $ _____ _____

5. side = 4.5 yd.

$A = $ _____ _____

6. side = 5.4 cm

$A = $ _____ _____

7. $s = \frac{1}{4}$ in.

$A = $ _____ _____

8. $s = 2\frac{1}{2}$ ft.

$A = $ _____ _____

9. $s = \frac{3}{4}$ ft.

$A = $ _____ _____

Area of a Triangle with a Formula

The **height** of a triangle is the perpendicular distance between the base and its opposite vertex. The **base** and height of each triangle are indicated.

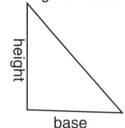 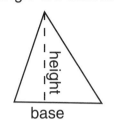

The **area** of a triangle is equal to $\frac{1}{2}$ the area of the rectangle around it.

$$A = \tfrac{1}{2} \text{ of (length} \times \text{width)}$$
$$= \tfrac{1}{2} \text{ of (base} \times \text{height)}$$
$$= \tfrac{1}{2}\, b\, h$$

Draw a rectangle around each triangle so that the area of a triangle is $\frac{1}{2}$ the area of the rectangle.

1.

Area of rectangle = _____

Area of triangle = _____

2.

Area of rectangle = _____

Area of triangle = _____

3.

Area of rectangle = _____

Area of triangle = _____

Find the area. Write the formula. Substitute and solve.

4.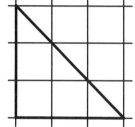

A = _____ (formula)

= _____ (substitute)

= _____ square units

5.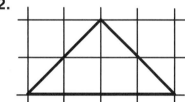

A = _____

= _____

= _____ square units

6.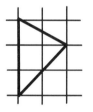

A = _____

= _____

= _____ square units

7.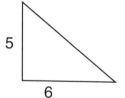

A = _____

= _____

= _____ square units

8.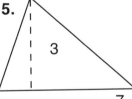

A = _____

= _____

= _____ square units

9.

A = _____

= _____

= _____ square units

Area of a Parallelogram with a Formula

Kari, John and Xi's group were asked to find the area of the parallelogram shaded on graph paper.

height

base

Kari cut off a corner part of the parallelogram and placed it to complete a 4 × 3 rectangle.

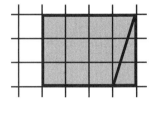

Area of rectangle = length × width = 12

Area of parallelogram = base × height = 12

Formula: Area = bh

Outline a rectangle having the same area as each shaded parallelogram.

1.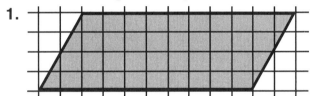

2.

Find the area. Write the formula. Substitute and solve.

3.

8 cm
11 cm

$A =$ _____ (formula)

$=$ _____ sq. cm

4.

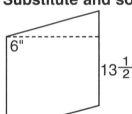

$6''$
$13\frac{1}{2}''$

$A =$ _____

$=$ _____ sq. in.

5.

$8'$
$5'$

$A =$ _____

$=$ _____ sq. ft.

6.

7.7 m
3.3 m

$A =$ _____

$=$ _____ sq. m

7.

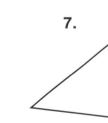

$7'$
$13'$

$A =$ _____

$=$ _____ sq. ft.

8.

6 in.
2 ft.

$A =$ _____

$=$ _____ sq. ft.

John found another way of finding the area of the parallelogram by thinking of two triangles inside the parallelogram. He drew this line.   *Can you explain John's method?*

Area of a Circle

You can find the area of a circle with the help of your old friend, pi (π).

A circle with radius r has a square drawn around it. We say that the circle is inscribed in the square.

The large square is divided into 4 smaller squares, each with a side of r.

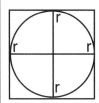

Area of small square = r^2
Area of large square = $4r^2$
Area of circle = __ r^2

Can you guess what number goes here?

Pi, the special relationship found in circles, is the missing number.

$$A = 3.14 \times r^2$$
$$A = \pi r^2$$

I can see that 3.14 is reasonable because the area of the circle is less than $4r^2$ but more than $3r^2$.

Find the area of a circle if the radius is 5 cm.

5 cm

$A = \pi r^2$
$= 3.14 \times (5)^2$
$= 3.14 \times (5 \times 5)$
$= 78.5$ square cm

Find the area of each circle. Write the formula, substitute and solve.

1.

2 in.

A = _____ (formula)
= _____ (substitute)
= _____ square in.

2.

3 ft.

A = _____ (formula)
= _____ (substitute)
= _____ square ft.

3.

2.1 cm

A = _____ (formula)
= _____ (substitute)
= _____ square cm

4.

1.2 km

A = _____ (formula)
= _____ (substitute)
= _____ square km

Area from Coordinate Grids

You can use a coordinate plane to draw geometric figures. You can find the perimeter and area of the figure from the drawing.

1. Plot the following points.

$A = (-3, 3)$ $B = (3, 3)$
$C = (3, -3)$ $D = (-3, -3)$

Figure ABCD is a _____.

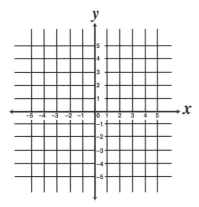

Length for each side = _____ units
Perimeter = _____ units
Area = _____ square units

2. Plot the following points.

$A = (2, 1)$ $B = (2, 4)$ $C = (5, 4)$

Figure ABC is a _____.

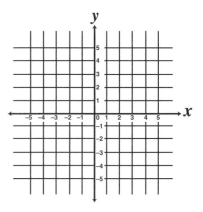

Area = _____ square units

3. Plot the following points.

$L = (-5, 5)$ $M = (0, 5)$
$N = (2, 3)$ $O = (-3, 3)$

Figure LMNO is a _____.

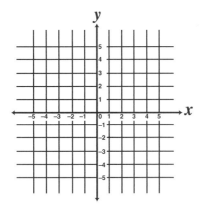

Area = _____ square units

4. Plot the following points.

$P = (-2, 3)$ $Q = (2, 3)$
$R = (2, -2)$ $S = (-2, -2)$

Figure PQRS is a _____.

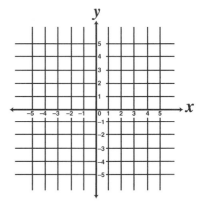

Area = _____ square units

Volume of a Rectangular Solid with a Formula

This rectangular solid has 2 layers (height). Each layer is 4 units long (length) and 3 units wide (width).

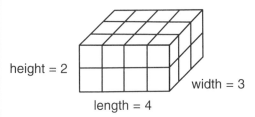

height = 2
width = 3
length = 4

How many cubic units?

One layer will have (4x3) 12 cubic units. There are 2 layers. . . There are 24 cubic units in all.

Volume = length × width × height

Formula: $V = l\,w\,h$

$$= 4 \times 3 \times 2$$
$$= 24 \text{ cubic units}$$

Write the formula, substitute and solve to find the volume of each solid.

1.

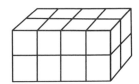

$V =$ _____ (formula)
$V =$ _____ (substitute)
$V =$ _____ cubic units

2.

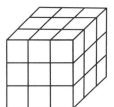

$V =$ _____
$V =$ _____
$V =$ _____ cubic units

3.

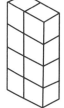

$V =$ _____
$V =$ _____
$V =$ _____ cubic units

4.

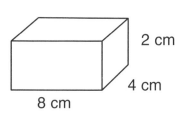

2 cm
4 cm
8 cm

$V =$ _____
$V =$ _____ cubic cm

5.

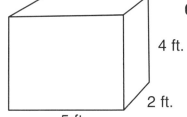

4 ft.
2 ft.
5 ft.

$V =$ _____
$V =$ _____ cubic ft.

6.

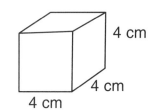

4 cm
4 cm
4 cm

$V =$ _____
$V =$ _____ cubic cm

7.

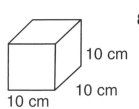

10 cm
10 cm
10 cm

$V =$ _____
$V =$ _____ cubic cm
$V =$ _____ cubic dm

8.

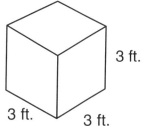

3 ft.
3 ft.
3 ft.

$V =$ _____
$V =$ _____ cubic ft.
$V =$ _____ cubic yd.

9.

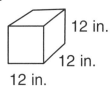

12 in.
12 in.
12 in.

$V =$ _____
$V =$ _____ cubic in.
$V =$ _____ cubic ft.

Volume of a Cube with a Formula

A **cube** is a special kind of rectangular solid in which the length, width and height are the same. The length, width and height are called the **edges** (e) of the cube.

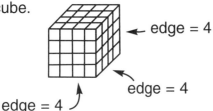

edge = 4

edge = 4

edge = 4

Instead of multiplying length x width x height, you can multiply edge x edge x edge.

Formula: $V = e \times e \times e$ or
$V = e^3$
$V = 4 \times 4 \times 4$
$= 64$ cubic units

1. $2^3 = 2 \times 2 \times 2 = $ _____

2. $5^3 = $ ___ $\times$ ___ $\times$ ___ $= $ _____

3. How are a rectangular solid and a cube alike? _____
_____

4. How are a rectangular solid and a cube different? _____
_____

Write the formula, substitute and solve to find the volume of each solid.

5.

$V = $ _____ (formula)
$V = $ _____ cubic units

6.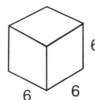

6

6   6

$V = $ _____
$V = $ _____ cubic units

7.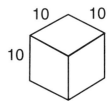

10   10

10

$V = $ _____
$V = $ _____ cubic units

Sketch each solid. Find the volume.

8. $l = 2$ m
$w = 3$ m
$h = 4$ m

$V = $ _____ cubic m

9. $l = 5$ in.
$w = 2$ in.
$h = 3$ in.

$V = $ _____ cubic in.

10. $e = 3$ ft.

$V = $ _____ cubic ft.

11. $e = 10$ cm

$V = $ _____ cubic cm

Volume of a Prism

A **prism** is a solid figure whose bases are congruent and parallel.
These solids are prisms and are named by the shape of their base.

| rectangular prism | triangular prism | trapezoidal prism | cylinder |

The volume of a prism is equal to the area of its base multiplied by its height.
B represents the area of the base in this formula.

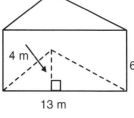

$V = Bh$ (A of $\triangle$ = $\frac{1}{2} bh$)

$= (\frac{1}{2} bh)h$

$= (\frac{1}{2} \cdot 13 \cdot 4)6$

$= 156$ cu m

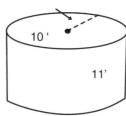

$V = Bh$ (A of $\bigcirc$ = πr^2)

$= \pi r^2 h$

$= 3.14 \times 100 \times 11$

$= 3454$ cu ft.

Which of these are prisms? _____

1.

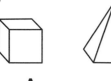 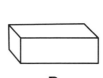

| A | B | C | D | E | F |

Find the volume of each prism. (Multiply the area of its base times its height).

2.

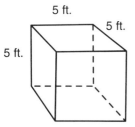

5 ft. 5 ft. 5 ft.

_____

3.

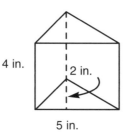

4 in. 2 in. 5 in.

_____

4.

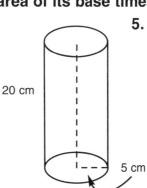

20 cm 5 cm

_____

5.

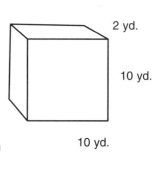

2 yd. 10 yd. 10 yd.

_____

How many cones filled with water will completely fill the drinking cup?

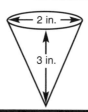

2 in. 3 in.

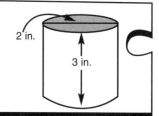

2 in. 3 in.

Using a Ruler to Find the Area of Irregularly Shaped Figures

Use a centimeter ruler to measure the length and width of each figure. Complete the chart.

1.

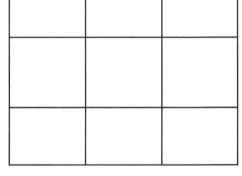

	length (cm)	width (cm)	area cm²
1.			
2.			
3.			
4.			
5.			
6.			

2.

3.

4. **5.** **6.**

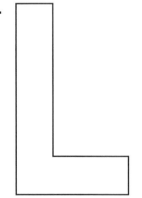

Area of Irregularly Shaped Figures

Formulas For Area

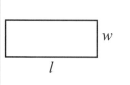

$A = lw$

$A = s^2$

$A = \frac{1}{2}bh$

$A = bh$

$A = \pi r^2$
($\pi = 3.14$ or $3\frac{1}{7}$)

Use the above formulas to find the area of each figure.

1.

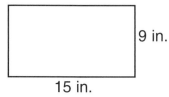

9 in.

15 in.

Area = _____ sq. in.

2.

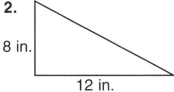

8 in.

12 in.

Area = _____ sq. in.

3.

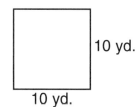

10 yd.

10 yd.

Area = _____ sq. yd.

4.

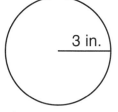

3 in.

Area = _____ sq. in.

5.

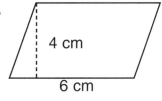

4 cm

6 cm

Area = _____ sq. cm

6.

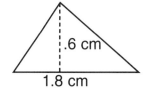

.6 cm

1.8 cm

Area = _____ sq. cm

7.

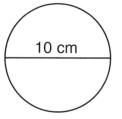

10 cm

Area = _____ sq. cm

8.

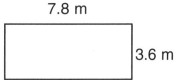

7.8 m

3.6 m

Area = _____ sq. m

9.

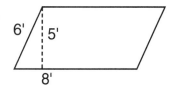

6'   5'

8'

Area = _____ sq. ft.

Can you find the area of each shaded figure?

A.

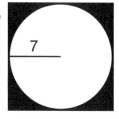

7

B.

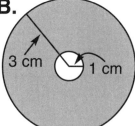

3 cm   1 cm

C.

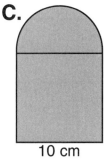

10 cm

10 cm

D.

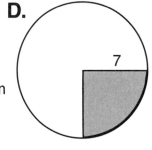

7

Area in the Metric System

The square units in the metric system are developed from the linear units (units used to measure lines). Metric square units come from metric line segments.

1 cm

1 cm | 1 sq. cm | 1 cm
1 cm

Area is the number of square units needed to cover a surface. How many square centimeters (cm^2) will be needed to cover this figure?

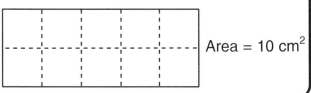

Area = 10 cm^2

How many square centimeters will be needed to cover each surface?

1.

Area = _____ cm^2

2.

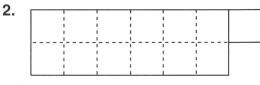

Area = _____ cm^2

3. Use centimeter graph paper to outline a square decimeter.

1 sq. decimeter = ____ sq. centimeters

4. Make a model of a 1 meter square by taping together or outlining 100 decimeter squares.

1 sq. meter = _____ sq. decimeters
1 sq. meter = _____ sq. centimeters

Estimate the area of each surface to the nearest square unit indicated. Compare your estimate with others in the class. Measure each surface with square units and compare your measurements to your estimate.

5. a postage stamp
estimate ____ sq. cm actual ____ sq. cm

6. a quarter sheet of this paper
estimate ____ sq. cm actual ____ sq. cm

7. a half sheet of this paper
estimate ____ sq. cm actual ____ sq. cm

8. the light switch by the door
estimate ____ sq. cm actual ____ sq. cm

9. the top of your desk or table
estimate ____ sq. cm actual ____ sq. cm

10. this sheet of paper
estimate ____ sq. cm actual ____ sq. cm

11. the smallest window in the room
estimate ____ sq. cm actual ____ sq. cm

12. the ceiling in the room
estimate ____ sq. cm actual ____ sq. cm

13. 300 sq. cm = _____ sq. dm

14. 450 cm^2 = _____ dm^2

15. 2 sq. m = _____ sq. dm

16. 325 dm^2 = _____ m^2

Area in the Customary System

The square units used for measuring area are named by the length of the line segment on each side of the square.

1 inch

1 inch

1 inch | 1 square inch | 1 inch

1 inch

Area is the number of square units needed to cover a surface. How many square inches will be needed to cover this figure?

Area = 2 square inches

How many square inches will be needed to cover each surface?

1.

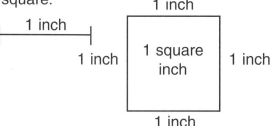

Area _____ sq. in.

2.

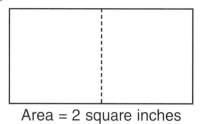

Area _____ sq. in.

3. Draw and measure a 1-foot square on a large sheet of paper. Draw lines to divide your large square into 1-inch squares.

1 square foot = _____ square inches

4. Make a model of a 1-yard square by taping together 1-foot squares.

1 square yard = _____ square feet

Estimate the area of each surface to the nearest square unit indicated. Compare your estimate with others in the class. Measure each surface with square units and compare your measurement to your estimate.

5. This sheet of paper

est _____ sq. in.   actual _____ sq. in.

6. Your desk or table

est. _____ sq. in.   actual _____ sq. in.

7. The largest window in your room

est. _____ sq. ft.   actual _____ sq. ft.

8. The floor of your room

est. ___ sq. yd.   actual _____ sq. yd.

9. 2 sq. ft. = _____ sq. in.

10. 4 sq. yd. = _____ sq. ft.

11. 288 sq. in. = _____ sq. ft.

12. 27 sq. ft. = _____ sq. yd.

13. A living room has 180 sq. ft. of floor space. How many square yards of carpeting will be needed to cover the floor?

_____ sq. yd.

14. A roll of wallpaper contains 36 sq. ft. How many rolls will be needed to wallpaper 4 walls if each wall needs 180 sq. ft.?

_____ rolls

Surface Area

The surface area of a cube is the sum of the areas of its faces.

Find the least amount of paper needed to cover a tissue box that is 5 cm × 5 cm × 5 cm.

How many faces does a cube have? How many square centimeters are there on one face of the cube?

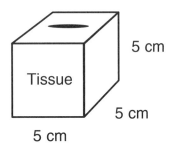

5 cm
Tissue
5 cm
5 cm

A net

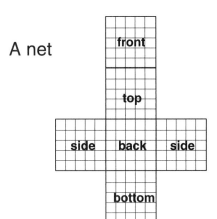

front
top
side   back   side
bottom

The area of each face = _____ sq. cm

The surface area of the cube = _____ sq. cm

Formula for surface area of a cube = _____

Centimeter graph paper can be folded over the box to make a net or pattern.

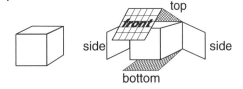

top
front
side   side
bottom

Find the surface area of each cube.

1.

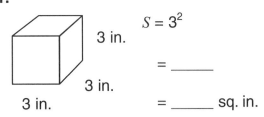

3 in.
3 in.
3 in.

$S = 3^2$

= _____

= _____ sq. in.

2.

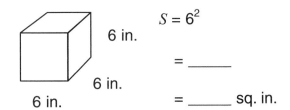

6 in.
6 in.
6 in.

$S = 6^2$

= _____

= _____ sq. in.

3. What happens to the surface area of a cube when an edge is doubled? What is the ratio of the surface areas? (Hint: try edge = 1 and then edge = 2)

4. What happens to the surface area of a cube when an edge is tripled? What is the ratio of the surface areas? (Hint: try edge = 1 and then edge = 3)

Volume in the Customary System

The cubic units for measuring the capacity an object will hold (its volume) are developed from line segments of equal length. The line segments become the edges of the cube.

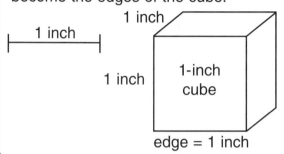

1 inch

1 inch

1 inch

1 inch

1-inch cube

edge = 1 inch

One cubic inch (cu in.) is the space inside a 1-inch cube.

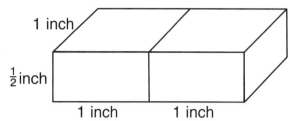

1 inch

$\frac{1}{2}$ inch

1 inch   1 inch

Is this a model of a 1-inch cube? Explain.
Is this a model of 1 cubic inch? Explain.

If ▱ = 1 cubic inch (not actual size), how many cubic inches will be needed to fill each box?

1.

volume = _____ cu in.

2.

volume = _____ cu in.

3.

volume = _____ cu in.

4. Make a model of a 1-inch cube from 1-inch graph paper. Use your model to answer the questions.

number of edges = _____

each edge = _____ in.

volume = _____ cu in.

5. Make a model of a 1-foot cube from large sheets of cardboard. Use your model to answer the questions.

number of edges = _____

each edge = ___ ft. = ___ in.

1 cu ft.= _____ cu in.

6. Make a model of a 1-yard cube from the sides of a large cardboard box. Use your model to answer the questions.

number of edges = _____

each edge = ___ yd. = ___ ft.

1 cu yd. = _____ cu ft.

Which cubic unit (inch, foot, yard) would be most appropriate to measure the volume of:

7. the box on a pickup truck

_____

8. a tissue box

_____

9. a shoebox

_____

10. an aquarium

_____

11. a swimming pool

_____

12. a refrigerator

_____

13. 2 cu. ft. = _____ cu in.

14. 2 cu. yd. = _____ cu ft.

15. $1\frac{1}{2}$ cu. ft. = _____ cu in.

16. 36 cu. ft. = _____ cu yd.

17. 864 cu. in. = _____ cu ft.

18. $1\frac{1}{3}$ cu. yd. = _____ cu ft.

Volume in the Metric System

The volume of this cube is 1 cubic centimeter (cm³). The capacity of a cubic centimeter is 1 milliliter.

1 centimeter cube has edges of 1 cm.

A milliliter is used to measure small quantities and is often used in medicine. An eyedropper holds about 1 milliliter (mL).

1 mL

The volume of this cube is 1 cubic decimeter (not actual size). The capacity of a cubic decimeter is 1 liter (L). A liter is a little larger than a quart. A medium bottle of soda holds 1 liter (L).

1 liter

SODA

(not actual size)

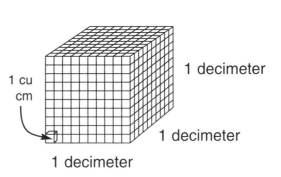

1 cu cm

1 decimeter

1 decimeter

1 decimeter

If [cube] = 1 cu cm (not actual size), how many cu cm will be needed to fill each box?

1.

volume = _____ cu cm

2.

volume = _____ cm³

3.

volume = _____ cm³

4. Make a model of a 1 cm cube from 1 centimeter graph paper. Use your model to answer the questions.
number of edges = _____
each edge = _____ cm
volume = _____ cm³ = _____ mL

5. Make a model of a 1 dm cube from centimeter graph paper. Use your model to answer the questions.
number of edges = _____
each edge = _____ dm
volume = _____ dm³ = _____ L = _____ cm³

6. 2 cu dm = _____ cu cm

7. 1 liter = _____ milliliters

8. 2000 cu cm = _____ cu dm

9. 40 mL = _____ cm³

Which cubic unit (centimeter or decimeter) would be most appropriate to measure the capacity of:

10. a bar of bath soap

11. a bathtub

12. a spice can

13. your desk drawer

14. thimble

15. a box of sugar cubes

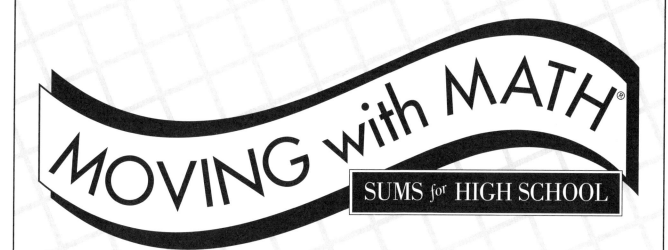

Success Using Math Standards for High School:
Preparation for Success

Part II

Math Teachers Press, Inc.

A Complete
Program for
Math Success in
High School

Probability, Statistics and Data Analysis

Table of Contents

Probability as a Ratio

Probability is another name for chance. The chance of something happening may be expressed as a ratio comparing the number of favorable outcomes to the number of possible outcomes.

You have a fair penny. What is the probability of flipping the coin and showing "heads" up?

There are two possible outcomes: heads or tails. There is one favorable outcome.

$$\text{Probability} = \frac{\text{number of favorable outcomes}}{\text{number of possible outcomes}} \qquad \text{Probability} = \frac{1}{2}$$

1. You flip a coin 100 times. In theory, how many times will heads show up?

_____

2. You flip a fair penny 5 times. Each time it lands heads up. What is the probability it will land on heads the next flip?

_____

You have a spinner divided into four fair parts.

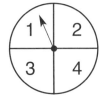

3. What is the probability the spinner will land on 3?

_____

4. What is the probability the spinner will land on 3 or 4?

_____

5. What is the probability the spinner will land on an even number?

_____

6. What is the probability the spinner will land on a number greater than 1?

_____

Probability as a Decimal or Percent

The probability of an event occurring may be expressed as a ratio, decimal or percent.

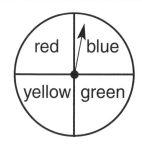

Probability = $\dfrac{\text{Favorable outcomes}}{\text{Possible outcomes}}$

Ratio $= \dfrac{1}{4}$

What is the probability the spinner will land on blue?

Express the probability as a ratio, decimal and as a percent.

To change $\dfrac{1}{4}$ to a decimal, think:

4 times what number is 100?

$\dfrac{1}{4} \overset{\times 25}{\underset{\times 25}{}} = \dfrac{25}{100} = 0.25 = 25\%$

Express each probability as a ratio, decimal and percent.

1. What is the probability that the above spinner will land on red?

_____ _____ _____

2. What is the probability that the spinner will land on red or blue?

_____ _____ _____

You throw a fair ten–sided die. One of the numbers 0–9 is written on each side of the die.

3. What is the probability the die will land with the "6" showing?

_____ _____ _____

4. What is the probability the die will land with an odd number showing?

_____ _____ _____

5. What is the probability that the die will land with a number greater than "4" showing?

_____ _____ _____

6. What is the probability that the die will land with a "0" or a "5" showing?

_____ _____ _____

Probability of a Single Event

Probability is another name for "What's the chance of this event happening?"
The probability of something happening will range between 0 and 1.

You have a bag of black marbles. You reach in and pull out 1 marble. What is the probability of pulling out a white marble?

Probability of 0 – impossible event

You have a dime and you flip it. What is the probability it will land on "heads"?

$$P = \frac{\text{desired outcome}}{\text{possible outcomes}}$$

There are 2 possibilities – heads or tails.

$$P = \frac{1}{2}$$

You have a bag of black marbles. You pull out 1 marble. What is the probability of pulling out a black marble?

Probability of 1 – a certain event

1. Use the spinner at the right to give the probability of spinning and . . .

   a. landing on 3? _____

   b. landing on 8? _____

   c. landing on an even number? _____

   d. landing on 9? _____

   e. landing on a number above 6? _____

   f. landing on a 1, 2, 3 or 4? _____

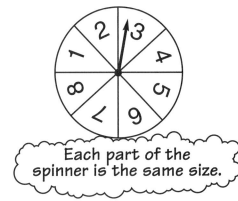

Each part of the spinner is the same size.

2. A box contains 11 cards. Each card has a letter from the word "PROBABILITY" written on it. What is the probability of drawing a card with "A"? "L"? "B"?

   _____ _____ _____

3. A die may show a 1, 2, 3, 4, 5, or 6 on its face. What is the probability of showing a 6? An even number? A 1?

   _____ _____ _____

4. A deck of cards has 4 suits of 13 cards each. The suits are called "clubs," "diamonds," "hearts" and "spades." What is the probability of drawing a spade? A club or spade?

   _____ _____

5. A bag contains 5 red, 5 blue and 10 white marbles. What is the probability of choosing a white marble? A red marble? A red, white or blue marble? A black marble?

   _____ _____ _____ _____

Probability of an Event Not Occurring

A die has six faces. A number from one to six is written on each face. If the die is thrown once, what is the probability of a 6 appearing? What is the probability of a 6 <u>not</u> appearing?

Probability $= \dfrac{\text{favorable outcomes}}{\text{possible outcomes}}$

$P(6) = \dfrac{1}{6}$

There are 5 outcomes that are <u>not</u> a "6". (1, 2, 3, 4, 5)

$P(\text{not } 6) = \dfrac{5}{6}$

The probability of an event not occurring equals 1 minus the probability of an event happening.

$P(\text{not } 6) = 1 - \dfrac{1}{6}$

$= \dfrac{5}{6}$

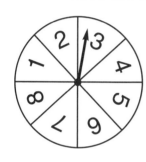

A spinner contains 8 equal parts. Each part is numbered from 1 to 8. If the pointer is spun once, what is the probability of:

1. The spinner landing on 4? _____

2. The spinner <u>not</u> landing on 4? _____

3. The spinner landing on a 7 or 8? _____

4. The spinner <u>not</u> landing on a 7 or 8? _____

5. The spinner landing on an even number? _____

6. The spinner <u>not</u> landing on an even number? _____

7. The spinner landing on a prime number? _____

8. The spinner landing on 9? _____

Probability and Prediction

A bag is filled with triangles, circles, squares and rectangles as shown below. If you reach into the bag 100 times and pull out a shape, replacing it each time, how many times do you predict you will pick each shape?

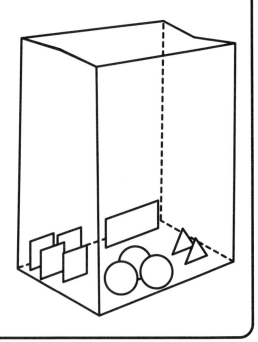

	1 pick	100 picks
Triangles	$\frac{2}{10} = \frac{1}{5}$	$\frac{1}{5}$ of 100 = 20
Circle	$\frac{3}{10}$	$\frac{3}{10}$ of 100 =
Square		
Rectangle		

P (△, ○, □ or ▭) = _____

Predict the number of favorable outcomes out of 1,000 spins.

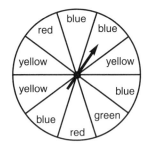

1. red   _____

2. not red   _____

3. blue   _____

4. not blue   _____

5. green   _____

6. blue or red   _____

7. Out of 100 light bulbs, 3 were found to be defective. How many would you expect to be defective out of 1000?

8. A computer randomly prints one-digit numbers from 1 through 5. If 100,000 of the numbers are printed, estimate the number of times the number "4" will occur.

_____

_____

A Fair Game

Game of Sums

Player A and B take turns throwing two 6-sided dice and adding the two numbers.

Player A gets 1 point if the sum is even.

Player B gets 1 point if the sum is odd. Would you like to be player A or B? Take turns throwing the dice 20 times to help you decide if the game is fair.

1. Tally Here

Even Sums	
Odd Sums	

2. Number of even sums? _____

Number of odd sums? _____

3. The game (is, is not) fair because _____

Table of Sums

+	1	2	3	4	5	6
1	2	3	4	5	6	7
2						
3						
4						
5						
6						

Game of Products

Player A and B take turns throwing two 6-sided dice and multiplying the two numbers.

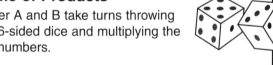

Player A gets 1 point if the product is even.
Player B gets 1 point if the product is odd.

Would you like to be player A or B?
Try 20 times with a partner to help you decide.

4. Tally Here

Even Products	
Odd Products	

5. Number of even products? _____

Number of odd products? _____

6. The game (is, is not) fair because _____

Table of Products

×	1	2	3	4	5	6
1	1	2	3	4	5	6
2						
3						
4						
5						
6						

A Tree Diagram of Outcomes

A **combination** is an arrangement in which the order is <u>not</u> important.

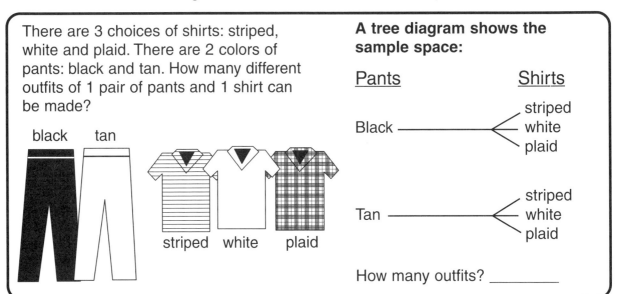

There are 3 choices of shirts: striped, white and plaid. There are 2 colors of pants: black and tan. How many different outfits of 1 pair of pants and 1 shirt can be made?

black   tan

striped   white   plaid

A tree diagram shows the sample space:

<u>Pants</u>   <u>Shirts</u>

Black ———————— striped
                        white
                        plaid

Tan —————————— striped
                        white
                        plaid

How many outfits? _____

1. You toss 3 different coins: a quarter, a dime and a nickel. How many different outcomes are possible?

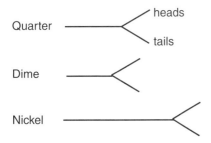

Quarter ———————— heads
                         tails

Dime ————————<

Nickel ————————<

There are _____ different outcomes.

2. You select the "Lunch Special": a main course and vegetable. The main course is fish, chicken or beef. The vegetables are beans, corn or peas. How many different "Lunch Specials" are there?
Make a tree diagram to show the sample space.

There are _____ "Lunch Specials".

3. Study the number of different outcomes in the example and problems 1 and 2. Find a pattern for finding the number of different outcomes.

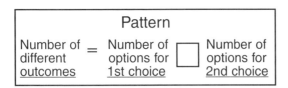

Pattern
Number of different <u>outcomes</u> = Number of options for <u>1st choice</u> ☐ Number of options for <u>2nd choice</u>

4. You have a combination lock of 3 numbers from 0 to 9. How many different combinations are possible?

Probability of Independent Events

Jami is going to spin three different spinners. Find the probability that all 3 spinners will land on "black". 	How many possible ways are there to spin the three spinners?

	When the first spinner is black. When the first spinner is white.
	1. B B B 5. W W W
	2. B B W 6. W W B
	3. B W W 7. W B W
	4. B W B 8. W B B

Probability = $\dfrac{\text{Favorable outcomes}}{\text{Possible outcomes}}$ Probability = $\dfrac{1}{8}$ *only 1 way is BBB* *8 different ways*	The multiplication principle: probability of separate events = product of each event. *Probability of spinning black each time.* $P = \dfrac{1}{2} \times \dfrac{1}{2} \times \dfrac{1}{2} = \dfrac{1}{8}$

1. Juan is spinning two black and white spinners as in the illustration. What is the probability that both spinners will land on white?

   _____

2. Bobbie is spinning two black and white spinners as in the illustration. What is the probability that neither will land on black?

   _____

3. How many different ways are there to flip a dime and throw a six-sided die? What is the probability of throwing "heads" and a "5" on the die?

   _____   _____

4. What is the probability of throwing two separate dimes and having them both land "heads" up?

   _____

5. What is the probability of throwing three separate pennies and having all of them land "tails" up?

   _____

6. How many different ways are there to throw two six-sided dice? What is the probability of throwing the dice so that both have a "4" showing?

   _____   _____

7. What is the probability of throwing "doubles" (matching numbers) when throwing two six-sided dice?

   _____

8. A multiple-choice test has 5 questions. Each question has 4 possible answers. What is the probability of "guessing" the correct answer to all 5 questions?

   _____

Independent and Dependent Events

Independent events: The outcome of previous events does not affect the outcome of events to follow. You have a red, blue and white block inside a bag. You pull a block and put it back before drawing a second block. What is your chance of pulling a red cube and then a white one?

1st pull 2nd pull

$$P = P(R) \times P(W)$$
$$= \frac{1}{3} \times \frac{1}{3} = \frac{1}{9}$$

Dependent events: The outcome of the first event affects the outcome of the second event. You have a red, blue and white block inside a bag. You pull a block and do <u>not</u> put the block back into the bag. You pull another block. What is your chance of pulling a red block and then a white one?

1st pull 2nd pull

$$P = P(R) \times P(W)$$
$$= \frac{1}{3} \times \frac{1}{2} = \frac{1}{6}$$

You have a bag containing 1 blue, 2 white and 3 red blocks.

1. What is the probability of pulling a red block and then a white block if you put the block back after your first pull?

    P = P(R) × **P**(W)

    = _____ × _____ = _____

    Independent or dependent events?

    _____

2. What is the probability of pulling a red block and then a white block if you do <u>not</u> put the block back after your first pull?

    P = P(R) × **P**(W)

    = _____ × _____ = _____

    Independent or dependent events?

    _____

You have a drawer with 2 black socks, 2 white socks and 2 gray socks.

3. What is the probability of pulling a black sock on one turn and a black sock on the next turn if you do <u>not</u> put the sock back after the first pull?

    P(BB) = **P**(B) × **P**(B)

    = _____ × _____ = _____

    Independent or dependent events?

    _____

4. What is the probability of pulling a black sock on one turn and a black sock on the next turn if you put the sock back after the first pull?

    P(BB) = **P**(B) × **P**(B)

    = _____ × _____ = _____

    Independent or dependent events?

    _____

Arrangements: Permutations

A **permutation** is an arrangement when the order is important.

Kari, John and Xi are singers in a talent show. In how many different orders can they sing?

Xi made a diagram of all the possible orders or arrangements.

	First	Second	Third
1.	X	K	J
2.	X	J	K
3.	K	X	J
4.	K	J	X
5.	J	X	K
6.	J	K	X

John had the singers stand in line in all different possible ways.

Try it. How many ways?

Kari thought about the problem.

There are 3 choices for the first singer. After the first singer has been chosen, there are 2 choices for the second singer. After the first and second singers are chosen, there is 1 choice for the third singer.

She said, "There are $3 \times 2 \times 1$ or 6 different orders."

1. Joy had a penny, nickel, dime and quarter. In how many different ways could she arrange them?

_____

2. Six students are seated in a row in a movie theater. In how many different orders can they be seated?

_____

3. How many ways can 5 people be posed in a line for a picture?

_____

4. How many different ways can 8 cheerleaders be arranged in a line?

_____

5. How many ways can the days of the week be arranged?

_____

6. There are 10 people on each team in a "tug-of-war" contest. How many different ways are there for people to line up?

_____

Finding an Average Number with Manipulatives

The **average** of a group of numbers is the number that comes closest to representing the values of all the other numbers. The average number will be greater than some of the numbers and less than others.

Jane has 5 pencils. The lengths of the pencils are 5 cm, 9 cm, 13 cm, 12 cm and 11 cm. What is the average length of the pencils?

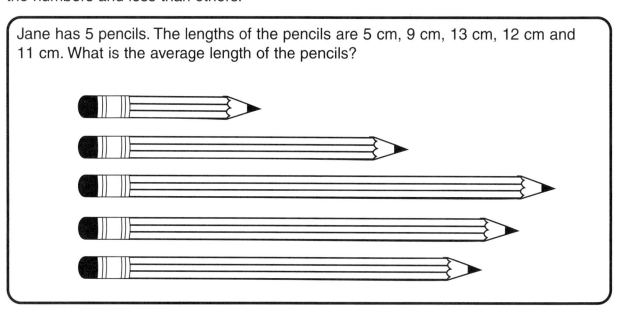

Find the average length of the following sets of pencils. Build each pencil with 1 centimeter unit blocks or squares. Adjust the pencil lengths until they are all equal.

1. Pencils of 5 cm, 9 cm, 13 cm, 12 cm and 11 cm.

   Average length: _____ cm

2. Pencils of 9 cm, 5 cm, 15 cm, and 11 cm.

   Average length: _____ cm

3. Pencils of 3 cm, 7 cm, and 2 cm.

   Average length: _____ cm

4. Pencils of 8 cm, 6 cm, 10 cm, 12 cm and 4 cm.

   Average length: _____ cm

5. Pencils of 12 cm, 8 cm, 9 cm, and 7 cm.

   Average length: _____ cm

6. Pencils of 18 cm, 12 cm, 14 cm, 13 cm and 23 cm.

   Average length: _____ cm

Average or Mean

Four children shared a box of cookies. The number of cookies each child received were: 6, 10, 11 and 9. Is there a fairer way to share the cookies?

> Put the cookies together and then pass them out, one at a time, to each child until there are none left.

6 + 10 + 11 + 9 = 36 cookies

$$\frac{9}{4\overline{)36}}$$ ←—The fair number of cookies is called the **average** or **mean**.

Use the same method to find the average of Andy's test scores.

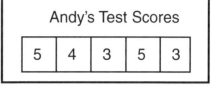

Andy's Test Scores

5	4	3	5	3

5 + 4 + 3 + 5 + 3 = 20

$$\frac{4}{5\overline{)20}}$$ ←— Andy's average test score

---

1. Susan scored 8, 5, 9, 6 and 7 on her tests. What was her average score?

   _____

2. In a week, Denise delivers 21, 17, 30, 24, 23, 25 and 21 papers. What is the average number she delivers each day?

   _____

3. In one week 24,731 people attended 7 baseball games. What was the average attendance per game?

   _____

4. Juan's spelling test scores were: 76, 80, 72 and 84. What was his average spelling score?

   _____

5. Doug scored 10, 16, 14 and 12 points in four basketball games. What was the mean number of points scored?

   _____

6. Vicki made an average of 87 on four tests. She knew that three of her scores were 82, 85 and 85. What was her fourth score?

   _____

7. The following distances were traveled by a car on an auto trip: 385 mi., 600 mi., 400 mi., 525 mi., 415 mi. and 375 mi. What was the average number of miles traveled each day?

   _____

8. A class raised money for a trip with car washes. On three Saturdays, they washed 90, 105 and 60 cars. What was the average number of cars washed each Saturday?

   _____

9. In a jump rope contest, Karla made the following number of jumps without missing: 30, 34, 36, 40 and 50. What was the average number of jumps?

   _____

10. Ed's father is 42 years old. Ed is 12 years old. His mother doesn't want to tell her age, but she tells Ed that the average age of the three family members is 30 years. How old is Ed's mother?

   _____

Collecting Data

Data Collected:_____

_____

SMALL GROUP DATA

Number in Group _____

Range of Data _____

Group Average _____

PREDICTION

Class Average _____

CLASS DATA

Number in Class _____

Range of Data _____

Class Average _____

Was your prediction reasonably close? Why or

why not? _____

_____

_____

Data Collected:_____

_____

SMALL GROUP DATA

Number in Group _____

Range of Data _____

Group Average _____

PREDICTION

Class Average _____

CLASS DATA

Number in Class _____

Range of Data _____

Class Average _____

Was your prediction reasonably close? Why or

why not? _____

_____

_____

Data Collected:_____

_____

SMALL GROUP DATA

Number in Group _____

Range of Data _____

Group Average _____

PREDICTION

Class Average _____

CLASS DATA

Number in Class _____

Range of Data _____

Class Average _____

Was your prediction reasonably close? Why or

why not? _____

_____

_____

Data Collected:_____

_____

SMALL GROUP DATA

Number in Group _____

Range of Data _____

Group Average _____

PREDICTION

Class Average _____

CLASS DATA

Number in Class _____

Range of Data _____

Class Average _____

Was your prediction reasonably close? Why or

why not? _____

_____

_____

Median, Extremes

A list of data is ordered from lowest to highest in numerical order. The highest and lowest numbers are called the **extremes**. The **median** is the middle number. If there are an even number of items, the median is the average of the two middle numbers.

Jane's math scores on five tests were: 75, 60, 55, 50 and 80. Find the extremes and the median of her test scores.

Order the numbers from least to greatest

Extremes
50   55   (60)   75   80
Median

José's math scores on six tests were: 84, 76, 60, 64, 72 and 74. Find the extremes and the median score.

← Extremes →
60   64   (72   74)   76   84

$Median = \frac{72 + 74}{2}$ or 73

The following table shows the height of the 15 students on the basketball team. Order the numbers from lowest to highest. Answer the related questions.

Student	Height in Inches	Student	Data Ordered
A	62	M	54
B	72		
C	68		
D	55		
E	74		
F	76		
G	74		
H	76		
I	78		
J	74		
K	60		
L	65		
M	54		
N	73		
O	77		

1. What are the extremes of the heights?

   _____ inches and _____ inches.

2. What is the median height?

   Player ____ with a height of _____ inches.

3. Which height was the most common? (the mode)

   _____ inches

4. Use a calculator to find the average height (mean) of the basketball players.

   _____ inches

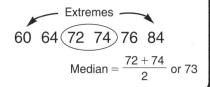

Mean, Median, Mode

A group of numbers can be represented by an average number. An average number is a central number that the other numbers cluster around. There are three kinds of average numbers - the **mean**, **median** and **mode**. You can use these bowling scores to help you understand each.

GAME SCORES
95	125	110	95	115

What is my bowling average?

The **mean** (often referred to as the average) is the arithmetical average of the numbers.

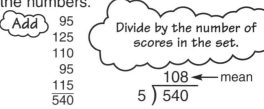

Add
95
125
110
95
115
540

Divide by the number of scores in the set.

$$5 \overline{)540} = 108 \leftarrow \text{mean}$$

The **median** is the middle number when the scores are ordered from greatest to least. The **range** is the difference between the highest and lowest score.

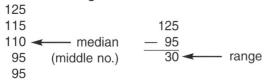

125
115
110 ←— median
95   (middle no.)
95

125
— 95
30 ←— range

The **mode** is the most popular number.

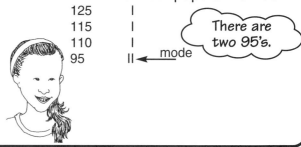

125 |
115 |
110 |
95 || ←— mode

There are two 95's.

---

1. Find the mean for Nate's test scores: 90, 84, 54, 78, 84.

_____

2. The heights of the five starters on the girl's basketball team are 68 in., 64 in., 70 in., 62 in., and 66 in. What is the average height?

_____

3. The scores of nine students in a math class were 74, 96, 75, 43, 100, 89, 62, 85 and 52. Find the range and median of the scores.

_____ _____

4. Tim drove 450 miles, 380 miles and 433 miles in a 3-day auto trip. What was his daily mean average?

_____

5. Find the mode of these temperatures:
a) 15°, 26°, 34°, 18°, 15° _____
b) 78°, 62°, 85°, 57°, 68° _____
c) 80°, 68°, 75°, 80°, 78° _____

6. The weights of five junior varsity football players are 72.4 kg, 80.3 kg, 62.8 kg, 67.5 kg and 92.5 kg. Find the mean weight.

_____

7.

	Complete data chart.	mean	median	mode	range
a	85, 65, 95, 100, 90, 95, 86				
b	180, 165, 160, 171, 173, 165				
c	120, 130, 160, 120, 150, 160				

Upper Quartile, Lower Quartile

Large sets of data are organized into four parts or **quartiles,** with each part having an equal number of **data points.**

The average temperature on 11 days were ordered from lowest to highest.
Draw a circle around the median temperature. Then draw a triangle around the data point for the upper and lower quartile.

65 67 68 69 70 71 73 75 78 79 80

        data point for   median   data point for
        lower quartile   temperature   upper quartile

Find the median, the data point for the upper quartile and the data point for the lower quartile of each set of data. Order the data when necessary.

	median	lower quartile point	upper quartile point
1. 7 8 9 10 11 13 14			
2. 19 24 16 25 30 26 20			
3. 54 72 67 64 77 73 66			
4. 9 5 1 2 4 1			

5. Your scores on eight math quizzes were: 70, 80, 74, 84, 76, 72, 73 and 82. What is the <u>upper</u> quartile point of your scores?

_____

(A) 80

(B) 81

(C) 82

(D) 84

6. Your scores on nine English tests were: 71, 82, 75, 73, 84, 70, 74, 76 and 80. What is the <u>lower</u> quartile point of your scores?

_____

(A) 70

(B) 71

(C) 72

(D) 74

Bar Graphs

Miltona Elementary School had an election for class president. This bar graph shows the results of the votes.

1. How many votes did each student receive?

   A Tony _____ **B** Marie _____

   C Tao _____ **D** Kay _____

2. What was the total number of votes?

   _____

3. How many fewer votes did Kay receive than Tao?

   _____

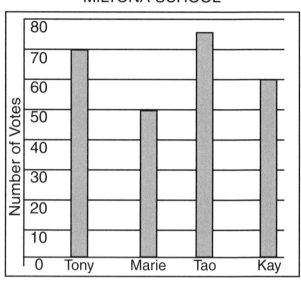

VOTES FOR PRESIDENT: MILTONA SCHOOL

Use the bar graph to answer these questions related to the enrollment at PS 24.

4. How many students were enrolled in 1998? 2001?

   _____   _____

5. How many more students were there in 1999 than in 2000?

   _____

6. Which year had the greatest enrollment? The least?

   _____   _____

ENROLLMENT AT PS 24

7. Predict what you expect the student population to be in 2002.

   _____

   _____

Line Graphs

Line graphs show changes over a period of time. Study the following graphs to answer the questions.

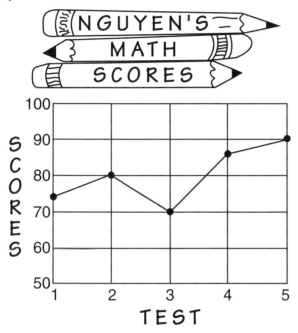

1. What is the highest score Nguyen received on the five math tests? The lowest? _____ _____

2. Estimate Nguyen's math scores for:
   Test 2 _____ Test 4 _____

3. How much higher was his score in week 5 than week 3? _____

4. Which is the best estimate of Nguyen's average score?

   A below 80 **B** about 80 **C** above 80

5. Find Nguyen's average score. _____

The graph illustrates the average temperature for each month in one year.

6. What was the average temperature in January? March?

   _____ _____

7. In what month does the highest average occur? The lowest?

   _____ _____

8. Which month shows the least increase in average temperature? _____

9. Between which two months was the greatest decrease in temperature?

   _____ and _____

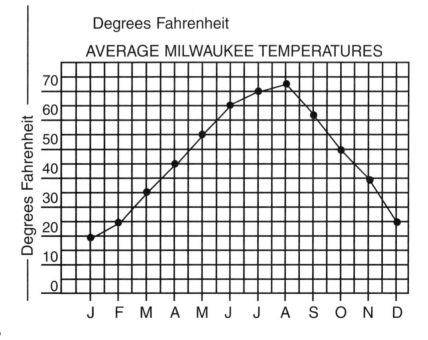

Circle Graphs

Circle graphs show how parts of a whole are related to the whole amount. The parts in a circle graph add to 100%.

Tina earned $3000 at her part-time job last year. The circle graph shows how she spent her money. How much did she spend on clothing?

40% of $3000 =
Solve this problem either by changing 40% to a fraction or a decimal.

As a fraction:

$\frac{40}{100} \times \$3000$

$\frac{2}{5} \times \$3000$

$1200.00

As a decimal:

$$\begin{array}{r} 3000 \\ \times \quad 40 \\ \hline \$1200.00 \end{array}$$

Both methods show that Tina spent $1200 on clothing.

How much did she spend on recreation? _____   Gifts? _____   Savings? _____

1. What item does Mike spend the most on? The least?   _____ _____

MIKE'S BUDGET

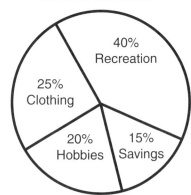

2. Mike earned $100 in September. How much did he spend on

   A recreation?   _____

   B clothing?   _____

   C hobbies?   _____

   D savings?   _____

3. Find the amount of money Mike would spend on each expense if he earned $200, $80, $150.

Amount Earned	$200	$80	$150
Recreation	a _____	e _____	i _____
Clothing	b _____	f _____	j _____
Hobbies	c _____	g _____	k _____
Savings	d _____	h _____	l _____

Reading a Bus Schedule

Mia has a job on Saturday from 9 to 3:30. She lives near 66th and Lyndale and takes the 47 bus to a downtown store near 7th Street.

Saturday

North: To Downtown								Dwntwn	
Bethany	Masonic Home	98th-Lyndale	86th-Lyndale	76th-Lyndale	66th-Lyndale	54th-Lyndale	35W and Lake	7th and 2Ave	Gateway Ramp
AM	AM	AM	AM	AM	AM	AM	AM	AM	AM
47C		653	705	710	715	719	725	732	739	744
47C		753	805	810	815	819	825	832	839	844
47C		853	905	910	915	919	925	932	939	944
47C		953	1005	1010	1015	1019	1025	1032	1039	1044
47C		1053	1105	1110	1115	1119	1125	1132	1139	1144
47C		1153	1205	1210	1215	1219	1225	1232	1239	1244
	PM	PM	PM	PM	PM	PM	PM	PM	PM	PM
47E	1250	1253	105	110	115	119	125	132	139	144
47C		153	205	210	215	219	225	232	239	244
47C		253	305	310	315	319	325	332	339	344
47C		353	405	410	415	419	425	432	439	444
47C		453	505	510	515	519	525	532	539	544
47X	650	653								

Dwntwn		South: From Downtown							
Gateway Ramp	Marq. And 8th	35W and Lake	54th-Lyndale	66th-Lyndale	76th-Lyndale	86th-Lyndale	98th-Lyndale	Masonic Home	Bethany
AM	AM	AM	AM	AM	AM	AM	AM	AM	AM
47C	752	800	807	814	820	824	829	834	846	
47C	852	900	907	914	920	924	929	934	946	
47C	952	1000	1007	1014	1020	1024	1029	1034	1046	
47C	1052	1100	1107	1114	1120	1124	1129	1134	1146	
47E	1152	1200	1207	1214	1220	1224	1229	1234	1246	1249
	PM	PM	PM	PM	PM	PM	PM	PM	PM	PM
47C	1252	100	107	114	120	124	129	134	146	
47C	152	200	207	214	220	224	229	234	246	
47C	252	300	307	314	320	324	329	334	346	
47C	352	400	407	414	420	424	429	434	446	
47C	452	500	507	514	520	524	529	534	546	
47E	552	600	607	614	620	624	629	634	646	649

SERVICE OPERATES MONDAY THROUGH SATURDAY EXCEPT ON THE FOLLOWING HOLIDAYS: New Year's, Memorial Day, Independence Day, Labor Day, Thanksgiving and Christmas

FARE INFORMATION

Adult Fare:	Off-Peak / *Peak
Rides completely within Zone 1............. $ .70	$.85

*Peak hours - Monday - Friday 6-9 AM and 3:30-6:30 PM.

1. What time should she meet her bus in order to be on time for work? _____

2. What number bus should she look for? _____

3. What time will she arrive at 7th and 2nd Ave.? _____

4. How long will she be on the bus? _____

5. What will her ticket cost? _____

6. On her return home she can catch her bus at the Gateway ramp or Marquette and 8th.

   What time will the first bus leave Marquette and 8th after work? _____

   What is the bus number? _____

   When will it arrive at 66th and Lyndale? _____

   If she misses that bus, how long will it be until the next bus? _____

7. How much will she spend for bus fare on Saturday? _____

8. Make up a question that could be answered from the bus schedule.

_____

_____

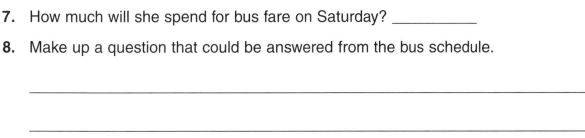

Box-and-Whisker Plots

Box-and-whisker plots summarize and display data.
The circled score is the middle or median score. The scores inside the triangles are data points for the upper and lower quartiles. The lower and upper extremes are underlined.

The scores of the average temperatures for 11 days are ordered from lowest to highest.

<u>65</u> 67 /68\ 69 70 (71) 73 75 /78\ 79 <u>80</u>

To make a box-and-whisker plot:

1. Plot the fine data points above a number line.

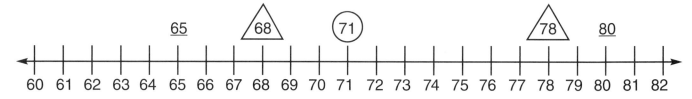

2. Draw a box enclosing the triangular points and a vertical line through the circled median score. Draw whiskers to the lower and upper extremes.

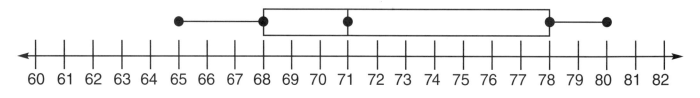

Draw a box-and-whisker plot for the following sets of test data. Remember to order and arrange the data above a number line.

Draw a line under the extreme scores, a circle around the median and triangles around the data points for the lower and upper quartiles.

1.   51 58 62 66 69 70 78 79 83 85 86

2.   65 68 70 82 87 88 88 92 98

Stem-and-Leaf Plots

A **stem-and-leaf plot** is a way to organize and display statistical data.

The average daily temperatures for 30 days are shown in the table.

60	70	79	40	61
58	67	71	54	83
67	61	60	54	61
85	75	63	51	68
78	66	63	62	69
64	75	86	43	69

These same temperatures may be organized into stem-and-leaf plots. In the left column, each digit in the tens place is listed. In the right column, the digits in the ones place are arranged from lowest to highest. The temperatures in the 40's and 50's have been entered. Enter the remaining temperatures.

stem	leaves
4	0 3
5	1 4 4 8
6	
7	
8	

Use the above data to complete the following problems.

1. The maximum (highest) average temperature was _____.

2. The minimum (lowest) average temperature was _____.

3. Most of the temperatures were in the _____

   A 40's
   B 50's
   C 60's
   D 70's
   E 80's

4. The median or middle score is _____.

   A 60
   B 61
   C 65
   D 67
   E 69

5. The mode temperature is _____.

6. The mean temperature is _____.

Scatter Plots

A **scatter plot** can be used to determine if a relationship exists between two sets of data. The data is plotted as sets of ordered pairs. In examining the plot, one can determine if there is a:

Positive Correlation – as one data set increases, so does the other.

No Correlation – the data sets are not related.

1. Listed below are shoe sizes and heights collected from males in an urban school. Graph and label the data on the coordinate grid.

Label	Shoe Size	Height
A	9	5 ft. 4 in.
B	11	6 ft. 2 in.
C	10	6 ft. 2 in.
D	8	5 ft. 9 in.
E	10	6 ft. 0 in.
F	9	5 ft. 8 in.
G	7	5 ft. 6 in.
H	11	5 ft. 11 in.
I	12	6 ft. 1 in.
J	8	5 ft. 7 in.
K	$9\frac{1}{2}$	5 ft. 8 in.

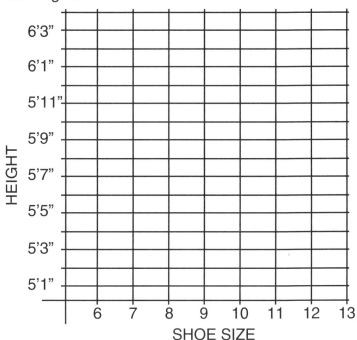

2. Draw a line that best fits the data. (DO NOT connect the dots. Draw a straight line whether it touches all of the points or not.)

3. Does the graph indicate a positive correlation between shoe size and height or is there no correlation? How do you know?

_____

_____

Unrelated Data Sets

When two sets of data are unrelated to each other, there will be no identifiable pattern in their scatter plot. The two data sets are **independent** of each other and have no correlation. Example: your shoe size and your annual salary.

1. Make a scatter plot of the information in the table on the coordinate grid provided.

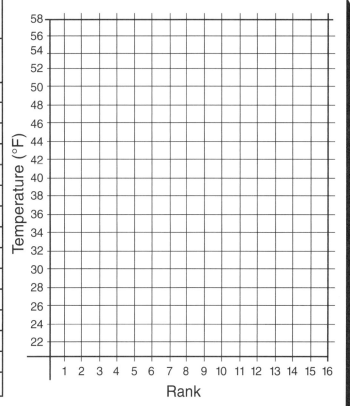

$$\$\$\$\$ \neq$$

Average January Temperatures of 15 Largest Cities		
Rank	City	Average Jan. Temperature
1	New York, NY	32°F
2	Los Angeles, CA	57°F
3	Chicago, IL	30°F
4	Houston, TX	50°F
5	Philadelphia, PA	31°F
6	Phoenix, AZ	54°F
7	San Diego, CA	57°F
8	Dallas, TX	45°F
9	Detroit, MI	23°F
10	San José, CA	50°F
11	Indianapolis, IN	26°F
12	San Francisco, CA	49°F
13	Jacksonville, FL	53°F
14	Columbus, OH	27°F
15	Austin, TX	49°F

2. Describe the relationship between the number of people that live in a given city and the average daily temperatures in January.

_____

3. The scatter plot has a (positive, negative or no) correlation. How can you tell?

_____

Will the pairs of information below have a positive, negative or no correlation?

4. Temperature and average number of people at the beach? _____

5. Average number of hours of sleep per day and height? _____

6. The amount of gas in a car and the number of miles already driven? _____

7. How fast you run and the size of your feet? _____

Probability, Data Analysis and Statistics Check Point

1. You spin a fair spinner divided into four parts. What is the probability the spinner will land on "C"?
(6SDAP 3.3)

2. You flip a fair coin three times. Each time the coin comes up heads. If you flip the coin again, what is the probability of heads showing?
(6SDAP 3.5)

3. You have pants in 3 different colors and shirts in 2 different colors. How many different outfits can you make?
(6SDAP 3.1)

4. The change in temperature for 5 days was 10°, 8°, 7°, 9° and 11°. What was the mean change in temperature?
(6SDAP 1.1)

5. The graph below shows the number of students that graduated from a high school over six years. Between which two years was there the greatest increase?
(6SDAP 2.5)

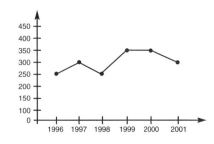

6. The boys basketball team has nine members. The heights of the members are 68 in., 72 in., 78 in., 70 in., 68 in., 70 in., 74 in., 76 in. and 71 in. What is the lower quartile data point of the heights?
(7SDAP 1.3)

7. The box-and-whisker plot shows times for running one mile at a track meet. What was the slowest time?
(7SDAP 1.1)

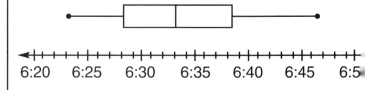

8. Mr. Mitchell's history class had the following scores on the last history test: 82, 97, 95, 100, 86, 81, 77, 65, 86, 92, 79, 72 and 64. Which is the correct stem-and-leaf plot for this data?
(7SDAP 1.1)

```
A    6 | 4 5
     7 | 2 7 9
     8 | 1 2 6 6
     9 | 2 5 7
    10 | 0
```

```
B    6 | 4 5
     7 | 2 7 9
     8 | 1 2 6
     9 | 2 5 7
    10 | 0
```

```
C    6 | 4
     7 | 2 7 9
     8 | 1 2 6 6
     9 | 2 5 7
    10 | 0
```

```
D    6 | 5 4
     7 | 7 9 2
     8 | 2 6 1 6
     9 | 7 5 2
    10 | 0
```

Algebra Functions and Algebra 1

Table of Contents

Adding Integers

Gregorio was scuba diving. He descended 30 meters below the surface and then ascended 5 meters. How far below the surface is he?

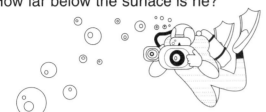

$-30 + 5 = -25$

Gregorio is 25 meters below the surface.

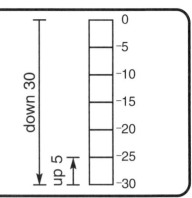

An elevator begins at ground level and makes the following trips. Complete the chart.

	Elevator goes	then goes	finishes at	equation
1.	up 5 floors	down 2 floors	+3	+5 + (−2)
2.	up 4 floors	down 5 floors		
3.	up 3 floors	down 3 floors		
4.	down 2 floors	up 5 floors		
5.	up 8 floors	down 11 floors		
6.	down 3 floors	up 3 floors		

Will each sum be positive or negative? Circle "+" or "−".

7. +20 + (−8)   (+ or −)   **8.** −14 + (+6)   (+ or −)

9. −7 + (+12)   (+ or −)   **10.** +10 + (−8)   (+ or −)

11. −53 + (+45)   (+ or −)   **12.** −17 + 0   (+ or −)

Find the sums.

13. +16 + (−9) = _____ **14.** −7 + (+7) = _____ **15.** 0 + (−14) = _____

16. +13 + (−25) = _____ **17.** +15 + (−25) = _____ **18.** −17 + (+8) = _____

19. −14 + (+8) = _____ **20.** −15 + (−6) = _____ **21.** −7 + (+15) = _____

22. +6 + (−12) = _____ **23.** +13 + 0 = _____ **24.** −9 + (+17) = _____

Can you find the pattern for the sign and number answer when you add a positive and negative number?

Subtracting Integers

The table shows the noon temperatures for 3 January days in Brainerd, Minnesota, and Minneapolis, Minnesota. What is the difference in temperatures between Brainerd and Minneapolis on each of the 3 days?

	Brainerd	Minneapolis
Monday	−5°	−3°
Tuesday	+5°	−3°
Wednesday	−5°	+3°

<u>Monday</u>
−5 − (−3) =

<u>Tuesday</u>
+5 − (−3) =

<u>Wednesday</u>
−5 − (+3) =

Start with 5 white cubes. Remove 3 white cubes.

−5 − (−3) = −2
Brainerd is __2° colder__.

Start with 5 black cubes. There are not enough negative cubes to take 3 away. Add 3 zero pairs.

+5 − (−3) = +8
Brainerd is_____.

Start with 5 white cubes. There are not enough positive cubes to take 3 away. Add 3 zero pairs.

−5 − (+3) = −8
Brainerd is_____.

Subtract:

1. +3 − (+6) = _____

2. +8 − (−3) = _____

3. +7 − (−2) = _____

4. −1 − (+8) = _____

5. +7 − (+4) = _____

6. +3 − (+6) = _____

7. +5 − (−7) = _____

8. −4 − (−5) = _____

9. +2 − (+1) = _____

10. +9 − (−4) = _____

11. +1 − (+10) = _____

12. −7 − (+10) = _____

13. −7 − (+8) = _____

14. −4 − (−4) = _____

15. −12 − (−15) = _____

16. +6 − (+4) = _____

17. −14 − (−14) = _____

18. +15 − (−22) = _____

19. Jacob's grandmother flew from Mexico City to Chicago. The temperature in Mexico City was 85°F. The temperature in Chicago was −20°F. What was the difference in temperatures?

_____

20. A plane took off and climbed 5000 feet and then descended 500 feet. How high off the ground was it after descending?

_____

Draw a picture and explain how to find the difference between a temperature of −6° and −8°.

Multiplying Integers

The product of two negative integers is positive.

Joan babysat for 3 hours. She earns $4 per hour. How much did she earn?

Put together 3 groups of 4 black cubes. This is multiplication.

The product of two positive integers is positive.

$3 \times 4 = 3 \cdot 4 = 3(4) = 12$

James is taking scuba diving lessons. The instructor says he will descend 5 feet every minute. How deep will he be in 3 minutes?

Put together 3 groups of 5 white cubes.

The product of a positive integer and a negative integer is negative.

$3 \times (-5) = -15$

The product of two negative integers is positive. $\quad -3 \times (-2)$

This expression means to subtract 3 groups of –2.

Start with 6 zero pairs.

Now remove 3 groups of –2.

$-3 \times (-2) = +6$

Find the products.

1. $-2 \times (+3) =$ _____ **2.** $-4 \times (-2) =$ _____ **3.** $-5 \times 2 =$ _____ **4.** $2 \times (-3) =$ _____

5. $4 \cdot 1 =$ _____ **6.** $-5 \cdot 6 =$ _____ **7.** $0 \cdot 7 =$ _____ **8.** $-3 \cdot -8 =$ _____

9. $-8\,(80) =$ _____ **10.** $5\,(-8) =$ _____ **11.** $+1 \times (+10) =$ _____ **12.** $-4\,(6) =$ _____

Complete the tables. Look for a pattern.

13.

Multiply by –3	
+3	
+2	
+1	
0	
–1	
–2	
–3	

14.

Multiply by –2	
+3	
+2	
+1	
0	
–1	
–2	
–3	

15. Positive or negative? The product of two integers with different signs is

_____.

The product of two integers with the same sign is _____.

Dividing Integers

You can find the quotient of two integers with different signs by relating a division fact to the related fact in multiplication.

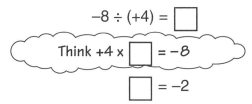

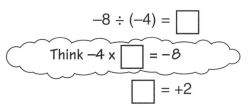

The quotient of two integers with different signs is_____.

The quotient of two integers with the same sign is_____.

Write the multiplication sentence for each division sentence. Solve for □.

1. $-6 \div 2 = \square$

$2 \times \square = -6$

$\square =$

2. $-12 \div 3 = \square$

$\square =$

3. $-9 \div 3 = \square$

$\square =$

4. $-15 \div 5 = \square$

$\square =$

5. $8 \div -2 = \square$

$\square =$

6. $15 \div -3 = \square$

$\square =$

7. $20 \div 5 = \square$

$\square =$

8. $16 \div -4 = \square$

$\square =$

9. $42 \div -6 = \square$

$\square =$

10. $-35 \div 7 = \square$

$\square =$

11. $64 \div 8 = \square$

$\square =$

12. $-81 \div 9 = \square$

$\square =$

13. $28 \div -7 = \square$

$\square =$

14. $56 \div -7 = \square$

$\square =$

15. The price of a stock fell $5 a day for 5 days. What was the total change?

$\square =$

16. When you multiply or divide two integers with the same sign, the answer is _____.

When you multiply or divide two integers with different signs, the answer is _____.

Draw a picture and explain how to solve $3 \times (-2) = x$.
Use the multiplication number sentence to write two related division sentences.

Variables in Algebraic Sums and Differences

A **variable** is a symbol that can represent an unknown number. A variable may be used with a plus or minus sign to show addition or subtraction.

Santo saves his pennies in a can he keeps on his dresser. How many pennies are in the can? 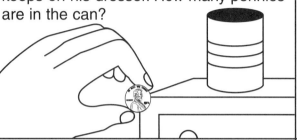	The number of pennies is unknown. The letter n can be used to stand for the unknown number. The letter n is called a variable. 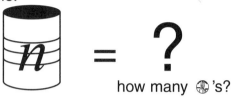 how many ⓟ's?
If Santo puts 4 pennies into the can, how many pennies would there be in the can? There would be $n + 4$ pennies.	If Santo takes 5 pennies from the can, how many pennies would be in the can? There would be $n - 5$ pennies.

Write an algebraic expression for each picture. Use n for the variable.

1. 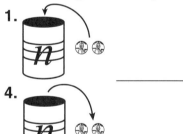 _____

2.  _____

3. _____

4. _____

5. _____

6.  _____

Write an expression with n for each phrase.

7. a number plus 4 _____

8. a number less 6 _____

9. a number decreased by 2 _____

10. a number increased by 7 _____

11. 20 plus a number n _____

12. a number minus 10 _____

13. 5 more than a number _____

14. 4 less than a number _____

15. 6 increased by a number _____

16. the sum of 9 and a number _____

17. a number minus 8 _____

18. difference between a number and 5 _____

19. 10 minus a number _____

20. the sum of 6 and a number _____

Variables in Algebraic Products or Quotients

A variable may be used in an algebraic expression to indicate multiplication or division.

Jami has two boxes of crackers. There are the same number of crackers in both boxes. How many crackers does Jami have?	Jami wishes to share one box of crackers on four plates. How many crackers will be on each plate?

Jami has two boxes of crackers. There are the same number of crackers in both boxes. How many crackers does Jami have?

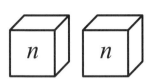

If n = number of crackers in one box,

$2 \times n$ = number of crackers in two boxes.

Note: $2 \times n$ is written $2n$ in algebra.

Jami wishes to share one box of crackers on four plates. How many crackers will be on each plate?

$$\frac{n}{4} = \bigcirc$$

If n = number of crackers in one box,

$4\overline{)n}$ = number of crackers on one plate.

n divided by 4 can be written:

$4\overline{)n}$ or $n \div 4$ or $\frac{n}{4}$

Write an expression for each picture. Use n for the variable.

1. _____

2. $\dfrac{\quad}{3}$ _____

3. $\dfrac{\quad}{2}$ _____

4. $\dfrac{\quad}{4}$ _____

Write an expression for the quotient for each picture.
Write your answer in 3 different ways.

5. $\dfrac{\quad}{3}$ _____ _____ _____

6. $\dfrac{\quad}{6}$ _____ _____ _____

Write an expression with n for each phrase.

7. 5 times a number _____

8. 8 groups of a number _____

9. a number divided by 3 _____

10. a number divided into 4 _____

11. 3 of a number _____

12. 12 divided by n _____

13. the product of n and 3 _____

14. the quotient of n and 2 _____

15. twice a number _____

16. the product of 6 and a number _____

17. a number divided by 10 _____

18. a number divided into 3 equal groups _____

Writing Expressions

Algebraic expressions may be pictured with rectangular rods and black and white cubes. The rectangular rods represent the unknown variable. The black and white cubes represent the positive and negative integers.

Example:    is written as: $3n + 4$

Example: is written as: $2n - 3$

Write the algebraic expressions for each picture. Use n to represent the variable.

1. _____

2. _____

3. _____

4. _____

Write the algebraic expressions for each picture. Use x for the unknown.

5. _____

6. _____

7. _____

8. _____

Write an algebraic expression with n for each word phrase.

9. six times a number increased by 2

10. twice a number less 4

11. six times a number plus the number

12. 3 times a number divided by 6

13. the sum of 3 times a number and 5

14. 4 times a number less 7

15. the difference of 2 times a number and 4

16. 5 groups of a number plus 2

17. 3 times a number divided by 4

18. 4 plus 3 times a number

19. the sum of a number, 2 times a number and 4

20. 2 times a number decreased by 4

Writing Equations from Models or Words

An **equation** is a sentence with an equal sign. In a mathematical sentence, the words "is equal to" are replaced by the equal sign. The equal sign tells you that both sides of the equation have the same value.

An equation can be written from a model or picture of a model.	An equation may be written from words.
 $x + 4 = 5$	The sum of three times a number and 4 is 10. *Sum means to add.* $3n + 4 = 10$

Write an equation for each picture. Use x as the variable.

1. $x - 3 = 4$

2. _____

3. _____

4. _____

5. _____

6. _____

7. A number plus 5 is 8.
$n + 5 = 8$

8. A number less 6 is 34.

9. 10 more than a number is 45.

10. A number increased by 6 is 58.

11. A number minus 20 is 80.

12. 2 times a number less 5 is 7.

13. 2 times a number plus 3 times a number is 30.

14. 3 times a number decreased by 4 is 23.

15. A number divided by 2 is 12.

16. 11 less than a number is 7.

17. Twice the number is 12.

18. 4 times a number is 40.

19. A number decreased by 4 is 50.

20. The quotient of 15 divided by a number is 3.

What's my number? Can you find the numbers that make problems 7–20 true?

Combining Like Terms

Joan has 1 full box of sugar cubes. Jackie has 2 full boxes of the same size and 5 more cubes. How many sugar cubes do they have in all? Joan ——Jackie——	Boxes containing the same number of objects can be represented by rods with the same dimensions. 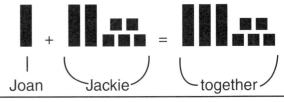 Joan Jackie together
You can use a variable to represent each rod or the number of cubes in each box.  x + $2x + 5$ Joan Jackie	Add the coefficients of the like terms. 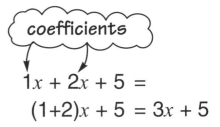 *coefficients* $1x + 2x + 5 =$ $(1+2)x + 5 = 3x + 5$

Use rectangular rods and black and white cubes to build the problems. Combine like terms. Write an equation showing the value of the rods and cubes before and after simplifying like terms.

1.

_____ = _____

2.

_____ = _____

3.

_____ = _____

4.

_____ = _____

5.

_____ = _____

6.

_____ = _____

7.

_____ = _____

8.

_____ = _____

Can you find a pattern for combining like terms?

Combining Like Terms Involving Addition

An algebraic term has two parts — a coefficient and a variable.

coefficient ⟶ $3x$ ⟵ variable

Like terms are made up of exactly the same variable and the same powers of these variables.

$-3x$, $4x$ and x are like terms.
7 and -2 are like terms.
x^2, and x are **not** like terms.
$2x$ and 2 are **not** like terms.

Like terms are simplified by combining the coefficients. Use the distributive property to combine terms.

Combine $2x$ and $1x$.
Combine 3 and 1.

$(2x + 3) + (x + 1)$
$2x + x + 3 + 1$
$(2 + 1)x + 4$
$3x + 4$

Write the coefficient for each term.

1. $6x$ _____   2. $4m$ _____   3. $-3x$ _____   4. n _____

Which pairs show like terms?

5. $-3c, 4c$ 6. $-2a, -2b$ 7. $3p, -p$ 8. $4n, 3n$ _____

List the like terms in each expression.

9. $3b + 4 + 6b$ 10. $2a - 7 - a$ 11. $5 + 4x - 3$ 12. $2n - 6 + 10$

_____   _____   _____   _____

Combine the like terms.

13. $3b + 4 + 6b$ 14. $2a - 7 - a$ 15. $5 + 4x - 3$ 16. $2n - 6 + 10$

_____   _____   _____   _____

Use rectangular rods and black and white cubes to combine like terms.

17.    _____

18.    _____

19. $(3x + 4) + (5x + 2)$ _____   20. $(n + 7) + (2n + 1)$ _____

21. $(4t + 6) + (3t - 2)$ _____   22. $(x - 2) + (3x + 5)$ _____

23. $(3a - 4) + (2a - 1)$ _____   24. $(x - 5) + (2x - 3)$ _____

25. $(4x + 1) + (x - 3)$ _____   26. $(2x + 3) + (x + 2)$ _____

27. $(2x - 1) + 3x$ _____   28. $(3x + 3) + (2x - 1)$ _____

29. $13 + 7c + (-4c) + 8$ _____   30. $4x + 7 + (-4) + (-x)$ _____

31. $6n + 7 - n$ _____   32. $12 + 6n + (-8) + (-4n)$ _____

Combining Like Terms Involving Subtraction

You can use rectangular rods and black and white cubes to subtract expressions. The difference may also be found by using the pattern for subtracting integers: add the opposite of each number being subtracted.

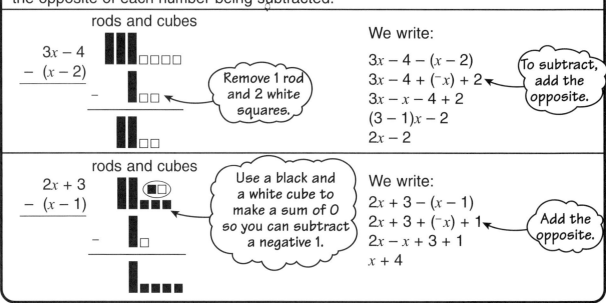

Use rectangular rods and black and white cubes to build each problem. Write the answer from the models.

1. $3x + 2$
   $- \ (x + 1)$

2. $2x - 4$
   $- \ (x - 1)$

3. $3x - 2$
   $- (2x + 1)$

4. $4n + 3$
   $- \ (n - 2)$

5. $2n - 2$
   $- \ (n - 3)$

6. $3n + 1$
   $- (2n + 2)$

Use the pattern of adding opposites to subtract each expression.

7. $5x - 4$
   $- (2x - 1)$

8. $4c + 3$
   $- (3c + 1)$

9. $2n - 6$
   $- \ (n + 2)$

10. $(x + 4) - (x + 2)$
    $x + 4 - x - 2$ = __2__

11. $(2n - 3) - (n - 1)$
    = _____

12. $(3a + 2) - (2a - 3)$
    = _____

13. $(4x - 1) - (2x + 3)$
    = _____

14. $(3x - 2) - (2x - 6)$
    = _____

15. $(5n + 4) - (2n - 3)$
    = _____

An Equation as a Balance

An **equation** is a mathematical sentence with an equal sign (=). The equal sign tells you that both sides of the equation should have the same value. An equation is like a balance scale where equal weights balance the scale.

A **true equation** will balance the scale.	A **false equation** will not balance the scale.	An **open equation** has a variable. To solve an open equation, you must find the value for the variable that will balance the scale.
		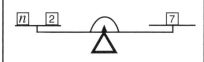
$4 + 3 = 7$ a true equation	$3 + 2 = 7$ a false equation	$n + 2 = 7$ an open equation

Write the equation shown in each balance.
Tell whether the equation is true, false or open.

1.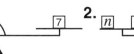

_____

_____

2.

_____

_____

3.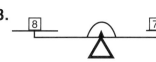

_____

_____

Write each equation. Give the value of the variable that will balance the scale.

4.

_____

$n =$ _____

5.

_____

$n =$ _____

6.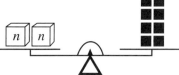

_____

$n =$ _____

7.

_____

$n =$ _____

8.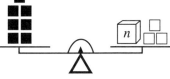

_____

$n =$ _____

9.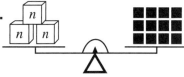

_____

$n =$ _____

Maintaining the Balance

An equation is like a teeter-totter. If a change is made on one side of the equation, the same change must be made on the other side for the equation to balance or be a true equation.

We can keep our balance if we make the same changes on both sides.

If I add 5 lb. to my side, you'll have to add 5 lb. to your side.

The two ends of a teeter-totter are balanced. Describe what change must be made to keep the balance even after:

1. 5 lb. is added on the right side.

_____

2. 10 grams is taken off the left side.

_____

3. The left side is doubled.

_____

4. The right side is divided by 2.

_____

Indicate what must be done to keep a true equation.

5. $9 + 4 = 9 + $ _____

6. $12 - 2 = 12 - $ ___

7. $6 \div 2 = 6 \div$ _____

8. $6 \times 2 = 6 \times$ _____

9. $n + 2 = 5$
$n + 2 - 2 = 5 - $ ___

10. $n - 3 = 7$
$n - 3 + 3 = 7 + $ ___

11. $2n = 6$
$2n \div 2 = 6 \div$ _____

12. $\dfrac{n}{4} = 2$

$4\left(\dfrac{n}{4}\right) = $ _____ $\times 2$

13. $m - 10 = 25$
$m - 10 + 10 = 25 + $ ___

14. $x + 15 = 40$
$x + 15 - 15 = 40 - $ ___

15. $3x = 12$
$3x \div 3 = 12 \div$ _____

16. $\dfrac{1}{2}n = 8$

$2 \times \dfrac{1}{2}n = $ ___ $\times 8$

What is the pattern for maintaining a true equation when an operational change (+, −, ✕ or ÷) is made on one side?

Solving Addition and Subtraction Equations with Models

To solve an equation, find the value for the variable which will balance the equation. You can use models to help find the solution.

Steps:	**Example:** $x + 3 = 8$	**Example:** $x - 2 = 5$
1. Build the equation with rods and cubes.		
2. Isolate x by using the sum zero pattern. Make the same change on the other side of the equation.	Add a ⁻3 to a ⁺3 to get a sum zero. Add a ⁻3 on the other side also. 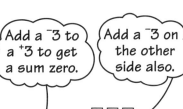	Add a ⁺2 to a ⁻2 to get a sum zero. Add a ⁺2 on this side also. 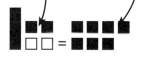
3. Simplify by matching positive and negative pairs.	$x = 5$	 $x = 7$

Use models to solve the equations.

1.

$x =$ _____

2.

$x =$ _____

3. $x + 3 = 6$

$x =$ _____

4. $x - 4 = 5$

$x =$ _____

5. $x - 2 = 6$

$x =$ _____

6. $x - 3 = 3$

$x =$ _____

7. $4 + x = 1$

$x =$ _____

8. ⁻4 + $x = 3$

$x =$ _____

9. $x - 4 = $ ⁻4

$x =$ _____

10. ⁻6 + $x = $ ⁻2

$x =$ _____

11. ⁻5 + $x = 3$

$x =$ _____

> *Can you find the pattern for solving equations with addition and subtraction?*

Properties of Equality

You can add or subtract the same value to each side of an equation to maintain equality. Addition and subtraction are opposites or inverse operations. One operation undoes the other: $4 + \mathbf{5} = 9$ and $9 - \mathbf{5} = 4$. To solve an addition or subtraction equation, use the inverse operation.

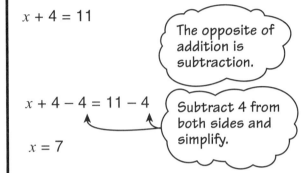

$x + 4 = 11$

The opposite of addition is subtraction.

$x + 4 - 4 = 11 - 4$

Subtract 4 from both sides and simplify.

$x = 7$

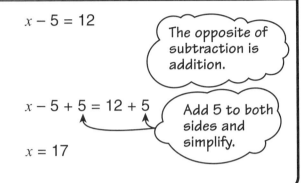

$x - 5 = 12$

The opposite of subtraction is addition.

$x - 5 + 5 = 12 + 5$

Add 5 to both sides and simplify.

$x = 17$

Solve the equations by adding or subtracting the same number to both sides.

1. $n + 2 = 5$

$\underline{n + 2 - 2 = 5 - 2}$

$n = \underline{\hspace{1cm}}$

2. $x - 4 = 5$

_____

$x = \underline{\hspace{1cm}}$

3. $n + 7 = 18$

_____

$n = \underline{\hspace{1cm}}$

4. $a - 8 = 12$

_____

$a = \underline{\hspace{1cm}}$

5. $x + 4 = 10$

_____

$x = \underline{\hspace{1cm}}$

6. $n - 14 = 10$

_____

$n = \underline{\hspace{1cm}}$

7. $x + 12 = 26$

_____

$x = \underline{\hspace{1cm}}$

8. $n - 24 = 30$

_____

$n = \underline{\hspace{1cm}}$

9. $n + 7 = 16$

_____

$n = \underline{\hspace{1cm}}$

10. $a + 4 = 17$

_____

$a = \underline{\hspace{1cm}}$

11. $x - 8 = 0$

_____

$x = \underline{\hspace{1cm}}$

12. $7 + x = 15$

_____

$x = \underline{\hspace{1cm}}$

Explain and draw pictures to show how you could use the property of equality to solve the equation $x - 6 = 2$.

Solving Multiplication Equations with Models

Multiplication and division are inverse operations. One operation undoes the other. You can divide both sides of an equation by the same number to maintain equality.

There are 12 flowers and 3 vases. If the same number of flowers are to go into each vase, how many flowers will go into each vase?

Let x = number of flowers in 1 vase.

$3x = 12$

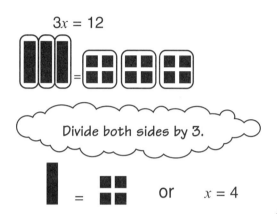

Divide both sides by 3.

or $x = 4$

Use models to solve the equations.

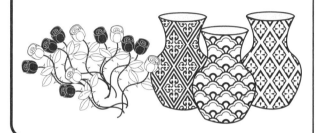

1. $x = $ _____

2. $x = $ _____

3. $2x = {}^-8$

   $x = $ _____

4. $5x = 10$

   $x = $ _____

5. $3x = {}^-6$

   $x = $ _____

6. $4x = 8$

   $x = $ _____

7. $4x = {}^-4$

   $x = $ _____

8. $2x = {}^-10$

   $x = $ _____

9. $3x = 9$

   $x = $ _____

10. $2x = {}^-6$

    $x = $ _____

11. $3x = 0$

    $x = $ _____

Jamie has $45. He has 3 times as much money as his sister. If x is the amount of money his sister has, write an equation that describes the money each has. Draw a picture and use words to explain how to find how much money his sister has.

Solving Division Equations with Models

You can multiply or divide both sides of an equation by the same number to maintain equality.

The boys in a physical education class were divided into six basketball teams. There were five playing on each team. How many boys were in the class?

Let b = the number of boys in the class.

$$\frac{b}{6} = 5$$

When the number of boys is divided into 6 groups, there are 5 in each group.

$$\frac{b}{6} \bullet 6 = 5 \bullet 6$$

$$\blacksquare = \odot \odot \odot \odot \odot \odot$$

$$b = 30$$

Use models to solve the equations. Draw pictures of your solutions. The first one has been done for you.

1. $\blacksquare \div 2 = 4$

$\blacksquare \div \frac{\bigcirc}{\bigcirc} = 4$

$\blacksquare = \bigodot \bigodot$

$\blacksquare = \underline{\ 8\ }$

2. $\blacksquare \div 3 = 4$

$\blacksquare = \underline{\qquad}$

3. $\blacksquare \div 5 = {}^-2$

$\blacksquare = \underline{\qquad}$

4. $\dfrac{\blacksquare}{3} = 6$

$\blacksquare = \underline{\qquad}$

5. $\dfrac{\blacksquare}{4} = 3$

$\blacksquare = \underline{\qquad}$

6. $\dfrac{\blacksquare}{2} = {}^-3$

$\blacksquare = \underline{\qquad}$

7. Molly had a bag of candy. It was shared between 6 people. Each person got 5 pieces. How many pieces were in the bag?

_____

8. Mr. Corrigan was having a math picnic. There were 23 students in his class. When the hot dogs were shared equally among 23 students, each student got 2 hot dogs. How many hot dogs did he buy?

_____

Solving Multiplication and Division Equations with the Inverse Operation

Multiplication and division are opposites or inverse operations. One operation undoes the other: $3 \times 4 = 12$ and $12 \div 4 = 3$. You can multiply or divide both sides of an equation by the same number and maintain equality.

Undoing Multiplication	Undoing Division
$4x = 20$	$\dfrac{x}{3} = 5$

The opposite of multiplication is division. Divide both sides by 4.

$$\frac{4x}{4} = \frac{20}{4}$$

$$x = 5$$

$$3\left(\frac{x}{3}\right) = (3)(5)$$

The opposite of division is multiplication. Multiply both sides by 3.

$$x = 15$$

Solve each equation by "undoing." Show your work.

1. $\dfrac{3n}{3} = \dfrac{21}{3}$

2. $\dfrac{n}{4} = 3$

3. $7x = 14$

4. $\dfrac{n}{6} = 4$

$n =$ _____ $n =$ _____ $x =$ _____ $n =$ _____

5. $4x = 168$

6. $\dfrac{n}{6} = 30$

7. $48 = 6n$

8. $\dfrac{c}{8} = 3$

$x =$ _____ $n =$ _____ $n =$ _____ $c =$ _____

9. $7x = 0$

10. $\dfrac{n}{3} = 7$

11. $9n = 180$

12. $20y = 180$

$x =$ _____ $n =$ _____ $n =$ _____ $y =$ _____

What's my number?

A $\dfrac{n}{0.4} = 1.2$ **B** $\dfrac{n}{\frac{1}{3}} = 15$ **C** $6n = 5.4$ **D** $\dfrac{3}{4}n = 9$

Solving Two-Step Equations

You can use rods and cubes to solve equations with two steps.

Alberto multiplied a number by 3 and added 4 to the product. The result was 10. What was the number?

By paper and pencil:

1. Build the equation.

$(3 \times \underline{}) + 4 = 10$

$3n + 4 = 10$

2. Subtract 4 from both sides.

$3n + 4 - 4 = 10 - 4$

$3n = 6$

3. Divide both sides by 3.

$n = 2$

With rods and cubes:

1. Build the equation.

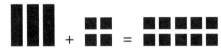

2. Add 4 white cubes to both sides.

Remove sums of zeros.

3. Divide both sides by 3.

Use models or draw pictures to solve each equation. Show your work.

1. $2x + 3 = 7$

$x = \underline{}$

2. $3n + (^-2) = 10$

$n = \underline{}$

3. $2n + 2 = 12$

$n = \underline{}$

4. $2n + (^-3) = 5$

$n = \underline{}$

5. $4x + (^-3) = 5$

$x = \underline{}$

6. $2n + 4 = (^-2)$

$n = \underline{}$

7. $3 + 2n = 9$

$n = \underline{}$

8. $2n - 4 = 2$

$n = \underline{}$

9. $2n + n + 1 = 7$

$n = \underline{}$

10. $\dfrac{n}{2} + 3 = 7$

$n = \underline{}$

11. $\dfrac{n}{3} + (^-1) = 1$

$n = \underline{}$

12. $\dfrac{n}{4} + 5 = 7$

$n = \underline{}$

13. The quotient of a number divided by 4 was increased by 3. The result was 7. What was the number?

\underline{}

14. The product of a number and 5 was decreased by 5. The result was 5. What was the number?

\underline{}

Two-Step Equations

You can solve equations with more than one step by using the properties of equality. First undo the addition and subtraction. Then undo the multiplication and division.

$2n - 5 = 7$ 1. Undo the addition or subtraction.

$2n - 5 = 7$
$2n - 5 + 5 = 7 + 5$

2. Undo the multiplication or division.

$2n = 12$
$$\frac{2n}{2} = \frac{12}{2}$$
$n = 6$

Solve.

1. $3n - 1 = 5$

$n = $ _____

2. $2n + 8 = 12$

$n = $ _____

3. $4x - 3 = 25$

$x = $ _____

4. $10n - 20 = 0$

$n = $ _____

5. $3 + 5n = 18$

$n = $ _____

6. $15 = 5x + 5$

$x = $ _____

7. $\dfrac{x}{4} + 3 = 5$

$x = $ _____

8. $\dfrac{x}{2} - 4 = 1$

$x = $ _____

9. $\dfrac{x}{4} + 4 = 7$

$x = $ _____

10. $\dfrac{n}{3} + 5 = 8$

$n = $ _____

11. $7 = \dfrac{n}{2} + 3$

$n = $ _____

12. $9 = 5 + \dfrac{n}{8}$

$n = $ _____

13. $16 = 4 + 3x$

$x = $ _____

14. $2 = \dfrac{n}{5} - 3$

$n = $ _____

15. $\dfrac{n}{10} - 1 = 2$

$n = $ _____

Combine like terms. Solve.

16. $3n + 2n + 5 = 20$

17. $5n - 8 + n = 4$

18. $3n - n + 7 = 13$

19. $8 = 5n + 2n - 6$

20. $4n + 3 - 2n - 6 = 5$

21. $7n - 4 + 9n = 28$

22. $5n - 4 + 2n = 17$

23. $14 + 2n + 5 = 39$

24. $2n + 8 + (-4) = 10$

25. Bernice said, "I am thinking of a number. If I multiply it by 3 and subtract 5 from the result, I get 40." What was Bernice's number?

26. Paul said, "I am thinking of a number. If I multiply it by 10 and add 5 to the result, I get 65." What was Paul's number?

Graphing the Solution for One Variable on a Number Line

After solving an equation involving one variable, the solution can be graphed on a number line.

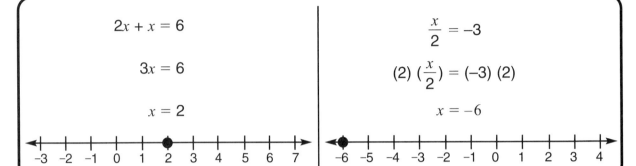

$$2x + x = 6$$

$$3x = 6$$

$$x = 2$$

$$\frac{x}{2} = -3$$

$$(2)\left(\frac{x}{2}\right) = (-3)(2)$$

$$x = -6$$

Solve for x. Check your solution. Graph the solution on the number line.

1. $x + 1 = 5$

$x =$ _____

2. $x - 4 = -2$

$x =$ _____

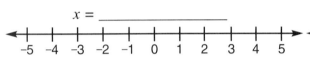

3. $\frac{x}{2} = 2$

$x =$ _____

4. $4x = -12$

$x =$ _____

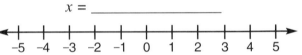

5. $2x + 3 = 7$

$x =$ _____

6. $\frac{x}{3} - 2 = -1$

$x =$ _____

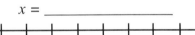

7. $2x - 4 = -10$

$x =$ _____

8. $\frac{x}{2} - 2 = 0$

$x =$ _____

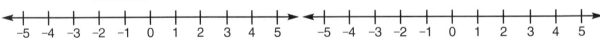

9. $|x| = 4$ $\quad$ (x can be +4 or −4.)

$x =$ _____

10. $|x| - 1 = 3$

$x =$ _____

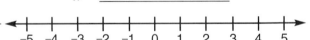
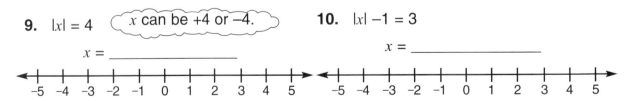

Inequalities

An equation is a sentence that contains an equal sign. A number sentence using the symbol *is greater than* (>) or *is less than* (<) is an **inequality.** An inequality can be true, false, or open.

True	$8 > 5$
False	$3 < -5$
Open	$n > 5$

More than one number can make an inequality true.

Which numbers make the inequality $n > 5$ true?

If $\{3, 4, 5, 6, 7\}$ is a set of possible solutions, only 6 and 7 make the inequality true.

Tell if each inequality is true, false, or open.

1. $2 < 6$ _____

2. $12 > 15$ _____

3. $x < -2$ _____

4. $4 > 5$ _____

5. $13 + 2 > 20$ _____

6. $x - 2 > 5$ _____

7. $n < 2$ _____

8. $x + 3 < 4$ _____

9. $4 \cdot 3 < 5$ _____

Write the symbol (> or <) that makes each sentence true.

10. $12 \bigcirc 10$

11. $7 \bigcirc 9$

12. $4 + 3 \bigcirc 8$

13. $6 - 2 \bigcirc 5 \cdot 2$

14. $3 + (-5) \bigcirc 1$

15. $2 + 4 \bigcirc -5$

Which number or numbers in each set make the inequality true?

16. $n < 4$ $\{0, 2, 4, 6\}$

17. $n > 3$ $\{-4, 0, 4\}$

18. $n + 2 > 5$ $\{0, 2, 4, 6\}$

_____

_____

_____

19. $n - 1 < 5$ $\{2, 4, 6, 8\}$

20. $3n > 9$ $\{2, 4, 6, 8\}$

21. $4n < 20$ $\{2, 4, 6, 8\}$

_____

_____

_____

22. $\dfrac{n}{2} > 3$ $\{2, 4, 6, 8\}$

23. $n + 7 > 12$ $\{2, 4, 6, 8\}$

24. $n - 3 > 2$ $\{2, 4, 6, 8\}$

_____

_____

_____

Solving Inequalities Involving Addition or Subtraction

You can use rectangular rods and black and white cubes to solve inequalities. Solve and graph.

$n + 3 > 5$

☐☐☐ █ ██ > ██ ███ ☐☐☐ ←

Add 3 white cubes to both sides.

█ > █ ← 2 is not a solution. The dot is not shaded.

$n > 2$

$n - 2 \leq 4$

≤ means less than or equal to

█ ☐ ██ ≤ ☐ ██ ██ ←

Add 2 black cubes to both sides.

█ ≤ ███ ← 6 is a solution. The dot is shaded.

$n \leq 6$

Number line: ⁻4 ⁻3 ⁻2 ⁻1 0 1 2 3 4 5 6 (open circle at 2, shaded to right)

Number line: ⁻2 ⁻1 0 1 2 3 4 5 6 7 8 (closed dot at 6, shaded to left)

Use rods and black and white cubes to solve each inequality. Graph the solution.

1.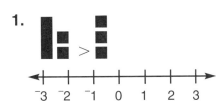

   Number line: ⁻3 ⁻2 ⁻1 0 1 2 3

2. █ ☐☐☐ < ██

3. $n - 6 > 4$

4. $n + 5 \geq 5$

5. $n - 3 < 7$

6. $n + 6 > 8$

7. $n + 4 > {}^-2$

8. $n - 3 \leq {}^-2$

9. The sum of a number n and 5 is less than 10. Write an inequality and solve for n.

10. A number n less 4 is greater than 2. Write an inequality and solve for n.

Solving Inequalities Involving Multiplication or Division

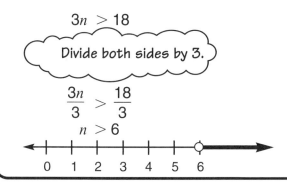

Divide or multiply both sides of the equation by the same numbers.

$3n > 18$

Divide both sides by 3.

$\dfrac{3n}{3} > \dfrac{18}{3}$

$n > 6$

```
0   1   2   3   4   5   6
```

$\dfrac{n}{5} \leq 4$

Multiply both sides by 5.

$(\dfrac{n}{5})(5) \leq (4)(5)$

$n \leq 20$

```
⁻10  ⁻5   0   5   10  15  20
```

Solve each inequality. Draw a picture of the solution on a number line.

1. $\dfrac{n}{2} < 5$ _____

2. $5n < 35$ _____

3. $3n > 27$ _____

4. $\dfrac{n}{4} > 8$ _____

5. $6x \geq {}^{-}12$ _____

6. $6x > 48$ _____

7. $\dfrac{x}{10} < 10$ _____

8. $\dfrac{x}{2} < {}^{-}3$ _____

9. $14x > 42$ _____

10. $\dfrac{x}{2} \geq 3$ _____

11. $\dfrac{x}{3} < 3$ _____

12. $\dfrac{x}{7} > 7$ _____

13. $5x > 0$ _____

14. $\dfrac{x}{10} > {}^{-}10$ _____

15. $8x \leq 16$ _____

Write an inequality using x. Solve for x.

16. Three times a number x is less than 18.

17. A number x divided by 2 is greater than 5.

18. Bill's brother is a salesperson in a sporting goods store. He earns 5% commission on all sales. He wants to earn no less than $125 to buy a car radio. Find how much he must sell.

19. The school band wants to give a concert. The auditorium rental will be no less than $600. The band members plan to sell concert tickets for $5.00 each. How many tickets must be sold to pay the rental?

Two-Step Inequalities

An inequality is solved the same way as an equation. If you add, subtract, multiply or divide both sides of an inequality by a positive number, the inequality remains the same.

Solve and graph the solution set for $3x + 4 > 10$.

Rods and Cubes

$3x + 4 > 10$

Add -4 (4 white cubes) to both sides.

$3x + 4 - 4 > 10 - 4$

Combine cubes to make pairs of zero.

$3x > 6$

Divide both sides by 3.

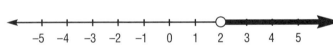

$x > 2$

Graph the solution set.

Solve and graph the solution set.

1. $5x - 2 < 13$

2. $2y - 4 > 8$

3. $7x - 4 > 3x + 8$

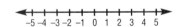

4. $\dfrac{x}{6} + 1 < 1$

5. $3x + 1 < 2(x - 1)$

6. $3a + 6 > a$

Solve. List 2 examples of numbers in the solution set.

7. $4x - 10 > 14$

8. $5x + 3 < 4x - 1$

9. $\dfrac{x}{3} + 5 > 7$

10. $7x + 4 < 3x - 8$

11. $\dfrac{x}{4} + 3 < 5$

12. $4x - 2 < 8 + 2x$

Two-Step Inequalities (≥ , ≤)

The less than or greater than symbols with lines under them are read: ≤ "less than or equal to" and ≥ "greater than or equal to." The graph of this inequality will have the endpoint shaded in because the endpoint number is included in the graph.

Solve and graph the solution set of $2x - 2 \leq 4$.

Rods and Cubes

Model the same way as an equation.

$2x - 2 \leq 4$

Add + 2 (2 black cubes) to both sides.
Join pairs of zero sums.

$2x - 2 + 2 \leq 4 + 2$
$2x \leq 6$

Divide both sides by 2.

$x \leq 3$

Graph the solution set. Shade in 3 because it is a member of the solution set.

```
←——+——+——+——◄——+——+——+——●——+——+——+——→
  -5  -4  -3  -2  -1   0   1   2   3   4   5
```

Solve. Graph the solution set.

1. What is the largest number that makes the statement true?

$x \leq 2$

2. What is the smallest number that makes the statement true?

$x \geq 7$

3. $x + 2 \geq 6$

4. $x - 4 \leq 7$

5. $3x - 1 \geq 2x + 2$

6. $4x - 5 \leq 7$

7. $\dfrac{x}{2} - 3 \leq -1$

8. $4x - 6 \geq 2x - 4$

9. $3x - 6 \geq x - 6$

10. $3x + 3 \leq x + 5$

11. $\dfrac{b}{3} + 7 \geq 9$

Two-Step Inequalities: Multiplication and Division by a Negative Number

An inequality is solved the same way as an equation.

The answer to an algebra problem should not have a negative sign in front of the variable.

$$-3x = 12$$

To solve −3 multiplied by x, we must do the opposite of multiplication. We must divide.

$$\frac{-3x}{-3} = \frac{12}{-3}$$

$$x = -4$$

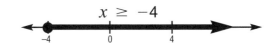

If you multiply or divide both sides of an **inequality** by a negative number, **reverse** the direction of the inequality symbol.

$$-3x \leq 12$$

Divide both sides by −3. Reverse the inequality.

$$\frac{-3x}{-3} \geq \frac{12}{-3}$$

$$x \geq -4$$

Solve the inequality for x. Graph the solution set on the number line.

1. $-x < 3$

2. $-2x > -2$

3. $\frac{-x}{2} \geq 1$

4. $-x + 4 \leq 2$

5. $-2x - 3 > -1$

6. $\frac{-x}{3} + 1 \geq 0$

Solve the inequality for x. Give two numbers in the solution set.

7. $2x + 3 \leq 4x + 5$

8. $x - 4 \geq 2x - 5$

9. $\frac{-x}{2} + 6 \geq -8$

10. $4x < 6x + 2$

11. $3x > 5x - 3$

12. $-3x + 4 \leq -5$

Parentheses Symbol

Parentheses symbols are used frequently in algebra. Parentheses must be followed carefully because they can indicate several actions. Here are three common actions.

Multiply	**Group the numbers** inside as a single value.	**"Do me first"**
(7)(–3) means 7 × –3 or 7 • –3 or –21	(8 + 7) means 15 (12 ÷ 2) means 6	5 + (4 – 2) 5 + 2 7 *(Do me first.)*

Do what the signs tell you to do. Write the answer.

1. (3)(4) _____

2. –2(6) _____

3. –5(4) _____

4. (5 + 3) _____

5. (12 + 6) _____

6. (17 – 9) _____

7. (4 • 8) _____

8. (49 ÷ 7) _____

9. (3 • 7) _____

10. 16 – (5 + 2) _____

11. (14 – 3) + 2 _____

12. (13 – 5) – 2 _____

13. (12 + 5) – 3 _____

14. (50 – 40) – 3 _____

15. 45 – (20 – 5) _____

16. (45 – 20) – 5 _____

17. (45 – 20) + 5 _____

18. 5 – (3 + 2) _____

19. 3 • (2 + 4) _____

20. (2 • 5) + 7 _____

21. (10 ÷ 5) – 1 _____

22. (8 – 3) × (4 – 2) _____

23. (80 ÷ 8) ÷ 2 _____

24. 3 + (12 ÷ 6) _____

25. (7 – 2) • (5 – 3) _____

26. (0 • 4) + 3 _____

27. (5 + 3)6 _____

Find the answers. How do the answers in A and B compare?

A. 25 – (10 + 5) = _____   **B.** (25 – 10) + 5 = _____

Can you explain the importance of carefully following the parentheses symbol?

Using the Distributive Property to Remove Parentheses

We know that multiplication is distributive over addition and subtraction. We often use this distributive property when doing "mental" arithmetic.

$$3 \times 29¢ = 3 \times (20 + 9)$$
$$= (3 \times 20) + (3 \times 9)$$
$$= 60 + 27$$
$$= 87¢$$

or

$$3 \times 29¢ = 3 \times (30 - 1)$$
$$= (3 \times 30) - (3 \times 1)$$
$$= 90 - 3$$
$$= 87¢$$

In algebra, the distributive property is used to remove parentheses.

with rods and cubes

$$3(n + 4) = 3 (\ \blacksquare \ \colon\colon \ )$$
$$= 3 (\ \blacksquare \ ) + 3 (\colon\colon)$$
$$= \blacksquare\blacksquare\blacksquare \ \colon\colon\colon\colon\colon\colon$$

we write

$$3 (n + 4) =$$
$$= 3(n) + 3(4)$$
$$= 3n + 12$$

Use rectangular rods and black and white cubes to remove parentheses. Draw a picture of your answer. The first one has been done for you.

1. $2(n + 3) = \underline{2 (\ \blacksquare \ \therefore \ )}$

$= \underline{\blacksquare\blacksquare \ \colon\colon\colon}$

$= \underline{\quad 2n + 6 \quad}$ Ans.

2. $3(n - 2) = \underline{\hspace{2cm}}$

$= \underline{\hspace{2cm}}$

$= \underline{\hspace{2cm}}$ Ans.

3. $2(n - 1) = \underline{\hspace{2cm}}$

$= \underline{\hspace{2cm}}$

$= \underline{\hspace{2cm}}$ Ans.

4. $4(n + 1) = \underline{\hspace{2cm}}$

$= \underline{\hspace{2cm}}$

$= \underline{\hspace{2cm}}$ Ans.

5. $2(2n - 1) = \underline{\hspace{2cm}}$

$= \underline{\hspace{2cm}}$

$= \underline{\hspace{2cm}}$ Ans.

6. $(2n + 3)2 = \underline{\hspace{2cm}}$

$= \underline{\hspace{2cm}}$

$= \underline{\hspace{2cm}}$ Ans.

Can you find the pattern for removing parentheses in each problem?

A. $3 \times (20 + 5)$ **B.** $2 \times (50 - 1)$ **C.** $a(b + c)$

Removing Parentheses and Combining Like Terms

Use the distributive property to remove parentheses.

Combine like terms.

$2\,(\ \blacksquare \vcenter{\hbox{$\cdots$}}\ ) + \blacksquare\blacksquare\blacksquare = 2(x + 4) + 3x$

$\blacksquare\blacksquare\vcenter{\hbox{$\cdots\cdots$}} + \blacksquare\blacksquare\blacksquare = 2x + 8 + 3x$

$\blacksquare\blacksquare\blacksquare\blacksquare\blacksquare\vcenter{\hbox{$\cdots\cdots$}} = 5x + 8$

Remove the parentheses and write the simplest name for each expression.

1. $5(a + 3)$ **2.** $4(n - 2)$ **3.** $5(b + 3)$ **4.** $6(x - 4)$

5. $7(2 + a)$ **6.** $3(4 - x)$ **7.** $9(3 - n)$ **8.** $8(b + 4)$

9. $4(2a + 3)$ **10.** $5(3b - 6)$ **11.** $2(3n - 4)$ **12.** $\frac{1}{3}(12x - 9)$

Remove the parentheses and simplify by combining like terms.

13. $2(x + 3) + x$ **14.** $5(x - 3) - 4x$ **15.** $7 + 2(x + 4)$

16. $3(x + 4) - 2$ **17.** $4(x - 3) + 10$ **18.** $4(2x - 3) + 20$

19. $2n + 4(n + 1)$ **20.** $3 + 2(5n - 6)$ **21.** $4(n + 5) - 7$

22. $2(7 - n) + 3n$ **23.** $3(n + 2) - n$ **24.** $4(n - 1) + (2)(3)$

25. $3(n + 2) - (2)(2)$ **26.** $2(n + 2) + 3(n - 1)$ **27.** $3(n - 1) - 2(n + 2)$

28. $\frac{1}{2}(6x + 10) - x$ **29.** $\frac{1}{3}(3x - 9) + 2x$ **30.** $\frac{1}{4}(20 - 8x) + 5x$

Order of Operations

The answer to an expression having mixed operations ($+, -, \times, \div$) will come out differently, depending on which operation you do first.

$$3 \bullet \underline{4 + 2}$$ **or** $$\underline{3 \bullet 4} + 2$$

First add the 4 + 2, then multiply the sum by 3.

First multiply the 3 • 4. Then add the 2.

$$3 \bullet 6 = 18$$ $$12 + 2 = 14$$

Do the operations in this order.

1. **M**ultiply and **D**ivide from left to right.
2. **A**dd and **S**ubtract from left to right.

$$16 \div 8 - 4$$
$$2 - 4 = -2$$

Divide before subtracting.

This rule is sometimes called the **M**y **D**ear **A**unt **S**ally rule (MDAS).

$$25 + (5)(2)$$
$$25 + 10 = 35$$

Multiply before adding.

Perform the operations in the correct order.

1. $3(4) - 6$ _____

2. $3 + (6)(4)$ _____

3. $13 + 4 - 2$ _____

4. $10 - 12 \div 4$ _____

5. $64 - 14(2)$ _____

6. $8 + (12)(-2)$ _____

7. $4(6) - 2 \bullet 12$ _____

8. $6 + 7 \bullet 2$ _____

9. $5 + 6 \div 3$ _____

10. $12 \div 4 - 2$ _____

11. $40 - 24 \div 4$ _____

12. $3 + 6 \bullet 5$ _____

13. $\dfrac{5(4)}{2}$ _____

14. $\dfrac{10}{5} + 8$ _____

15. $\dfrac{6(4)}{2(6)}$ _____

16. $\dfrac{10}{2} + 5 \bullet 3$ _____

17. $\dfrac{18}{9} + 2 \bullet 3$ _____

18. $\dfrac{81}{3(3)} - 5$ _____

19. $14 + 8(4) - 10$ _____

20. $70 - 15 \div 3$ _____

21. $10(6) - 12 \div 4$ _____

22. $6 \bullet 8 - 3 \bullet 4 + 2 \bullet 3$ _____

23. $30 - 20 \div 5 + 4$ _____

24. $3(20) - 10(3) + 5$ _____

Parentheses and the Order of Operations

When there are parentheses and mixed operations in an expression, always solve the operations in the parentheses before applying the correct order of operations. Remember the **M**y **D**ear **A**unt **S**ally rule: **M**ultiply, **D**ivide, **A**dd, **S**ubtract.

	Simplify	$18 - (3 + 2) + 4 \cdot 3$	Simplify	$5(7 + 3) - 12 \div 6$
1. Solve parentheses		$18 - 5 + 4 \cdot 3$		$5 \cdot 10 - 12 \div 6$
2. Multiply, divide		$18 - 5 + 12$		$50 - 2$
3. Add, subtract		25		48

1. $(15 - 9) + 4$

2. $40 - (16 + 2)$

3. $(40 - 16) + 12$

4. $7(9 - 4)$

5. $(8 + 7)3$

6. $64 - (14 + 2)$

7. $4(3 + 7)$

8. $(16 - 12)4$

9. $24 - 3(16 - 11)$

10. $3(6) + 5(10 + {}^-14)$

11. $2(40 - 30) - 7(2)$

12. $(3 + 6) \div 3 - 2$

13. $(7 + 4 + 7) \div 2 + 4$

14. $2(7 + 5) - (5 - 1)$

15. $16 \div 8 + 3 - 4$

16. $4(6 + 3) \div 9 - 4(0)$

17. $\frac{1}{2} (4 + 12)$

18. $\frac{1}{3} (10 - 1)$

19. $(7.4 - 1.4) \div 3$

20. $\frac{(3 + 6)}{(21 - 3)}$

21. $\frac{4(3)}{2} + 4$

22. $\frac{12}{6} + 8$

23. $\frac{100}{2(4 + 1)}$

24. $\frac{3(2 + 5)}{7}$

Insert parentheses so each expression names the greatest number.

A. $3 \times 4 + 9$

C. $32 \div 8 + 8$

E. $3 \times 10 - 6$

B. $16 + 10 \div 5$

D. $24 - 3 \times 2$

F. $5 \times 3 - 2$

Algebraic Terms

A **monomial** is a term like $4x$. A monomial may be a number (5), a variable (x) or a product of a number and one or more variables ($5n$, $3x^2$).

A **polynomial** is a sum of monomials.

$$r^2 + 2r - 5 \text{ is a polynomial.}$$

A **binomial** is a polynomial of two terms: $x^2 - 4$ or $2x + 3$.

A **trinomial** is a polynomial of three terms: $n^2 - 4n + 7$.

In the phrase $2x + 3x$, the items $2x$ and $3x$ are called **like terms. Like terms** are made up of exactly the same variable and the same powers of these variables.

Example: Simplify by adding similar or like terms.

$$2x + 4 + 3x - 1$$
$$2x + 3x + 4 - 1$$
$$(2 + 3)x + 3$$
$$5x + 3$$

2x and 3x are like terms. 4 and −1 are like terms.

1. Which expressions are monomials?

   a. $12x$ b. $x - y$ c. $4x^2$ d. $6ab$

2. Which expressions are binomials?

   a. $5x^3$ b. $a + b$ c. $5 - 7d$ d. $2x^2 + 4x - 3$

3. Which expressions are trinomials?

   a. $3a^2b^4$ b. $3a^2 + 6a + 3$ c. $x^2 + 14x + 49$ d. $5a \div 5$

Simplify each expression.

4. $4x + 2 + x - 1$ 5. $2a + 5 + a + 2$ 6. $2x + x + x - y$

7. $5a - 2b + 4 + 2a + b + 2$ 8. $4x + 5y + 6 + 3x + y$ 9. $r + 1 + 2r - 3$

10. $3x - (2x + 4) + 2$ 11. $(2x - 3) - 3(x + 2)$ 12. $3a + 4 - 2(a + 5)$

Multiplying Monomials Using the Commutative Property

Commutative Property	Monomials include a number, a variable or products of one or more variables. The parts of a monomial may be multiplied in any order.

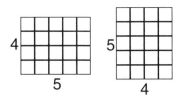

$$4 \times 5 = 5 \times 4$$
$$20 = 20$$

We can multiply in any order and the product is the same.

$$a \bullet b = b \bullet a$$

$a^2 \bullet a^3$

$a \bullet a \times a \bullet a \bullet a$

a^5

$(4a^2) \times (-2a)$

$4 \bullet a \bullet a \times (-2) \bullet a$

$(4)(-2) \bullet a \bullet a \bullet a$

$-8a^3$

First, find the sign of the product.   (+ or −)
Then find the numerical part.   (8)
Then find the variable part. (a^3)

Find the missing numbers or variables.

1. $3 \bullet 8 = $ _____ $ \bullet 3$

2. $18 \bullet 25 = $ _____ $ \bullet 18$

3. $2 \bullet b = b \bullet $ _____

4. $x \bullet y = y \bullet $ _____

5. $n^2 \bullet m = $ _____ $ \bullet n^2$

6. _____ $ \bullet x^4 = x^4 \bullet y$

Simplify.

7. $4x \bullet 3 = $ _____

8. $4a \bullet b = $ _____

9. $6c \bullet -cx = $ _____

10. $-7x \bullet 5 = $ _____

11. $-6x \bullet (-4) = $ _____

12. $-4a \bullet (-5a) = $ _____

13. $3^4 \bullet 3^6 = $ _____

14. $9^3 \bullet 9^2 = $ _____

15. $x^2 \bullet x^3 = $ _____

16. $-n^3 \bullet n^2 = $ _____

17. $-3a^2 \bullet a^4 b^2 = $ _____

18. $x^2 \bullet 2x^3 = $ _____

19. Complete the table for the powers of 3:

Powers of 3	3^1	3^2	3^3	3^4	3^5
Value	3	3(3) = 9	3(3)(3) = 27	3(3)(3)(3) = ____	3(3)(3)(3)(3) = ____

20. The product of exponents with a common base (i.e. $a^4 \bullet a^2$) is found by

_____ the exponents.

Multiplication of Monomials

The pattern for multiplying powers with the same base is suggested by the following multiplication problem.

100	×	100	=	10,000
↓		↓		↓
10 × 10	×	10 × 10	=	10 × 10 × 10 × 10
↓		↓		↓
10^2	×	10^2	=	10^4

You can multiply powers with the same base by adding exponents.

$$10^2 \cdot 10^2 = 10^{2+2} = 10^4$$

Product of Powers: For any number a and all integers m and n,
$$a^m \cdot a^n = a^{m+n}$$

Simplify.　　$(-3x^2y)(4xy^3)$

Reorder.　　$-3 \cdot 4 \cdot x^2 \cdot x \cdot y \cdot y^3$

$-12x^3y^4$

Determine the sign. Then multiply the numerical part. Then multiply like variables by adding exponents. Remember x means x^1.

Multiply these monomials:

1. $(4p^3q)(9p^2q^4)$

2. $(-7xy^4)(6x^3y^5)$

3. $(2st)(12s^5t^6)$

4. $(5rs^2)(13rs^2)$

5. $(-dr^2)(-6d^3)$

6. $(9x^4y)(-3y^6)$

7. $(8w^4y^3)(9wy)$

8. $(8c^4d^2)(-3cd)$

9. $(8abc^3)(-3a^2b^4c)$

10. $(7x^2yz)(4yz)$

11. $(3r^2s)(2r)(4s^4)$

12. $(3ab)(b^2c)(4bc^2)$

13. $(xy^2)(xyz)$

14. $(-p^4q^2)(2p^3a)$

15. $(-2mn^2)(-6m^2n^{-2})$

16. $(-6abc)(-6a^3bc^3)$

17. $(x^2y)(x^2y)(x^2y)$

18. $(ab)(ab)(ab)$

Powers of Monomials

An exponent tells how many times a number or variable is used as a factor.

$(2^3)^2 = 2 \bullet 2 \bullet 2 \bullet 2 \bullet 2 \bullet 2$
$= 2^{3+3}$ or 2^6

The outer exponent, 2, tells you 2^3 is used as a factor twice.

$(2x^3)^2 = (2)^2(x^3)^2$
$= 2 \bullet 2 \bullet x^3 \bullet x^3$
$= 4x^6$

Use $2x^3$ as a factor twice.

Power of a Power. For any number x, and all integers a and b,
$$(x^a)^b = x^{ab}$$

Evaluate the monomials below by removing the parentheses.

1. $(5x)^3 = \underline{5 \bullet 5 \bullet 5 \bullet x \bullet x \bullet x}$

   $= \underline{125\,x^3}$

2. $(4x)^2 = $ _____

   $= $ _____

3. $(2x)^3 = $ _____

   $= $ _____

4. $(7x)^2 = $ _____

   $= $ _____

5. $(9x)^2 = $ _____

   $= $ _____

6. $(10x)^3 = $ _____

   $= $ _____

7. $(8x)^2 = $ _____

   $= $ _____

8. $(6y)^2 = $ _____

   $= $ _____

9. $(11z)^2 = $ _____

   $= $ _____

10. $(1x)^5 = $ _____

    $= $ _____

11. $(4x^2)^3 = $ _____

    $= $ _____

12. $(2x)^5 = $ _____

    $= $ _____

13. $(2x^2y)^2 = $ _____

    $= $ _____

14. $(4yz)^2 = $ _____

    $= $ _____

15. $(3xy^2)^3 = $ _____

    $= $ _____

16. To find the power of a power, _____ the exponents.

Dividing Monomials

You can find the pattern for dividing monomials from these examples.

$$x^5 \div x^2 \quad \text{means} \quad \frac{\overset{1}{\cancel{x}} \cdot \overset{1}{\cancel{x}} \cdot x \cdot x \cdot x}{\underset{1}{\cancel{x}} \cdot \underset{1}{\cancel{x}}} = x^3$$

Cancel $\frac{x}{x}$ as 1.

Here is another way to solve the same problem:

We know that multiplication and division are inverse operations.

Since $x^2 \cdot x^3 = x^{2+3}$ or x^5, then $x^5 \div x^2 = x^{5-2}$ or x^3

Simplify $\dfrac{m^5 n^3}{m^2 n} = m^{5-2} n^{3-1} = m^3 n^2$

Remember this is n^1.

> Quotient of Powers. The exponent of any number or letter in a quotient is found by _____ the exponent of that letter in the divisor from the exponent of that letter in the dividend.
>
> For any non-zero number x and integers a and b,
>
> $$\frac{x^a}{x^b} = x^{a-b}$$

1. $\dfrac{7^3}{7^2}$

2. $\dfrac{8^7}{8^5}$

3. $\dfrac{a^{13} b^3}{a^6 b}$

4. $\dfrac{k^{16}}{k^{10}}$

5. $\dfrac{x^2 y^5}{x^3 y^2}$

6. $\dfrac{xy^5 z}{xy^2 z}$

7. $\dfrac{15^3}{15^2}$

8. $\dfrac{xyz}{xyz}$

9. $\dfrac{m^8 n^{12}}{m^5}$

10. $\dfrac{cx}{cy}$

11. $\dfrac{18a^4 n^5}{3a^3 n}$

12. $\dfrac{(3a)^2 b^4}{6ab}$

13. $\dfrac{n^4}{n^4} = n^{\square} =$ _____

14. $\dfrac{n^5}{n^5} = n^{\square} =$ _____

15. For any non-zero number, n, $n^0 =$ _____ .

Squares and Square Roots of Monomials

Squaring a number is the opposite of finding a square root.
The pair of operations act as inverses. They "undo" each other.

Kim was tiling the floor of the dog's pen. The pen is a square with 3 feet on each side. How many 1 foot squares will she need?

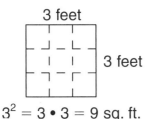

3 feet

3 feet

$3^2 = 3 \cdot 3 = 9$ sq. ft.

James bought 16 one-foot carpet squares. He wanted to glue them together to make a large square rug. How many feet are on each side of the rug?

4 ft. on a side
1	2	3	4
5	6	7	8
9	10	11	12
13	14	15	16

$\sqrt{16}$ means the square root of 16
$\sqrt{16} = 4$

To find the square root of a monomial with a variable,

1. Break into numbers and letters. $\sqrt{9x^4} = \sqrt{9} \ \sqrt{x^4}$
2. Find the square root of each part. $= 3 \quad x^2$

Think: What expression do you multiply by itself to get x^4?
$x^2 \cdot x^2 = x^4$

Find the square roots.

1. 16

2. 25

3. 4

4. 49

5. 81

6. 100

7. 121

8. 144

9. x^2

10. n^6

11. r^{10}

12. a^8

13. $\sqrt{16y^2}$

14. $\sqrt{m^2}$

15. $\sqrt{36n^6}$

16. $\sqrt{49g^2}$

17. $\sqrt{100x^8}$

18. $\sqrt{81}$

19. $\sqrt{9y^2}$

20. $\sqrt{4r^{10}}$

21. $\sqrt{4g^2h^2}$

22. $\sqrt{25x^2y^6}$

23. $\sqrt{64m^4}$

24. $\sqrt{16x^2y^2z^2}$

25. List all the numbers between 1 and 100 that have whole number square roots.

_____

_____

Polynomials and Their Opposites

You can use models of tiles to represent polynomials.

Polynomial Models	
Use three types of tiles to model polynomials.	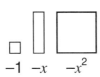

$x^2 + 3x - 4$

The opposite of
$x^2 + 3x - 4$ is : $-x^2 \quad -3x \quad + 4$

Each tile has an opposite.

The **opposite** or **additive inverse** of a polynomial is the polynomial which, when added to the original polynomial, results in 0. To find the opposite, just find the opposite of each term.

Find the opposite of $\quad 4x^2 + 3x - 2$

Solution: $\quad -4x^2 - 3x + 2$

Find the opposite of each term.

Represent these polynomials by drawing models.

1. $x + 4$

2. $2x^2 - 3x + 1$

3. $x^2 + 6x - 3$

4. $-3x^2 + 1$

5. $-2x + 3$

6. $2x^2 + x - 5$

Write the opposite or additive inverse of each polynomial.

7. $y - 7$

8. $-81z$

9. $-4c^2 + 14c - 7$

10. $-a + 4b - 5c$

11. $18m^2 - 9m - 45$

12. $x^2 - 4x$

13. $-3x^2 + 7x$

14. $2m^2 - 3m$

Addition of Polynomials, Degree Greater Than One

Method 1 $(x^2 + 3x + 1) + (2x^2 - 2x + 3)$

- Model each polynomial.
- Arrange like terms in columns.
- Remove zero pairs.
- Write the polynomial for the sum.

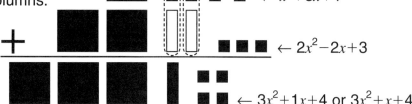

zero pairs

$\leftarrow x^2 + 3x + 1$

$\leftarrow 2x^2 - 2x + 3$

$\leftarrow 3x^2 + 1x + 4$ or $3x^2 + x + 4$

Method 2 $(x^2 + 3x + 1) + (2x^2 - 2x + 3)$

- Group like terms together. $[(1x^2) + (2x^2)] + [(3x) + (-2x)] + [(1) + (3)]$
- Combine like terms. $3x^2 + 1x + 4$ or $3x^2 + x + 4$

Method 3 $(x^2 + 3x + 1) + (2x^2 - 2x + 3)$

- Arrange like terms in columns. $(x^2 + 3x + 1)$
- Combine like terms.

$$\begin{array}{r} (x^2 + 3x + 1) \\ + \underline{(2x^2 - 2x + 3)} \\ 3x^2 + 1x + 4 \textbf{ or } 3x^2 + x + 4 \end{array}$$

Solve the problems. Be prepared to explain your method.

1. $(x^2 - 4x) + (3x^2 + 2x)$

2. $(-2m^2 + 3m) + (-7m - 2)$

3. $(3x^2 - 5x - 2) + (x^2 - x + 1)$

4. $(18y + 9) + (22y + 3)$

5. $(-13a + 9) + (8a - 9)$

6. $\begin{array}{r} -9x^2 - 6x - 3 \\ + \underline{5x^2 + 16x + 8} \end{array}$

7. $(2x^2 - 7x + 6) + (-3x^2 + 7x)$

8. $(6mn + 4m + 3) + (5mn)$

9. $(-x^2 + 6) + (3x)$

10. A triangular garden has sides as shown. Find the perimeter.

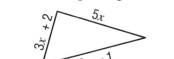

11. Find the perimeter of the figure.

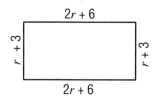

Perimeter means the distance around a figure. We can find the perimeter by adding together the measures of the sides.

Subtraction of Polynomials, Degree Greater Than One

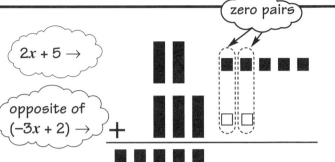

Method 1 $(2x + 5) - (-3x + 2)$

- Use the additive inverse to change the subtraction to addition of the opposite.
- Arrange like terms in columns.
- Remove zero pairs.
- Write the polynomial for the sum.

$2x + 5 \rightarrow$

opposite of $(-3x + 2) \rightarrow$

zero pairs

$\leftarrow 5x + 3$

Method 2 $(2x + 5) - (-3x + 2)$

additive inverse of $-3x + 2$ is $3x + -2$

- $(2x + 5) + (3x + -2)$
- Put like terms together. $(2x + 3x) + (5 - 2)$
- Simplify. $5x + 3$

Method 3 $(2x + 5) - (-3x + 2)$

- Find the additive inverse of $(-3x + 2)$ which is $(3x - 2)$.
- Arrange like terms in columns. $(2x + 5)$
- Combine like terms. $\underline{+ (3x - 2)}$
  $5x + 3$

Solve the problems. Be prepared to explain your method.

1. $(3x^2 + 7x) - (2x^2 + 2x)$

2. $(17a + 12) - (-8a - 2)$

3. $(11x^2 - 29x + 10) - (5x^2 - 13x - 20)$

4. $18y^2 - 2y + 2$
$\underline{-(9y^2 - 6y - 3)}$

5. $(8m^2 + 6m + 3) - (5m^2 - 2m)$

6. $(3x^2 + x) - (-4x^2 - 3x + 6)$

7. $(7mn^2 - 6m - n) - (3mn^2 - n)$

8. $(7c - 10d) - (3c + 4d)$

9. $(8y^2 + 4y + 8) - (6y^2 + 7y - 2)$

10. The perimeter of a triangle is $18m + 15$. Two of the sides added together equal $12m + 12$. What is the measure of the third side?

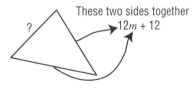

These two sides together
$12m + 12$

?

11. The expression for the total income of a business for one month was $13x^2 - 3x + 24$.
The expression for the costs for the month was $4x^2 - 6x - 7$.
Find the expression for that month's profit. (Total income − costs = profit.)

Multiplying a Polynomial by a Monomial

You have used rectangles to model multiplication. To visualize $x(x+2)$, think of a rectangle that is x units along one side and $x+2$ along the other. Use rods and squares to model $x(x+2)$.

The rectangle is made up of one x^2 tile and 2 x rods. The area of the rectangle is x^2+2x.

To multiply a polynomial by a monomial, multiply the monomial times each term of the polynomial: $x(x + 2) = x \cdot x + x \cdot 2 = x^2 + 2x$

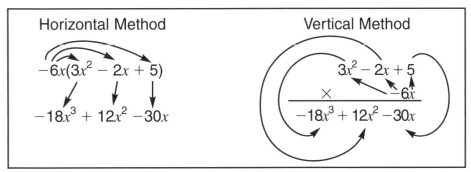

Multiply:

1. $4y(y - 6)$

2. $9t^2(t^2 + 3t - 3)$

3. $-9(3t^2 + 7t - 8)$

4. $-2v(v^4t + 6)$

5. $x^3yz(4x + 5y + 6z)$

6. $c^2d(5c - 43d + 9)$

7. $-k^3(2k + 4)$

8. $(-3x^2 - 5x - 7)x^3$

9. $20(2x^2 + x + 3)$

10. $-w^3(-2w^2 + 8w)$

11. $6t(4t^3 - 5)$

12. $7(-x^2 - 2x + 2)$

13. $8x(x^2 + 3x + 5)$

14. $-3(-4t - 9)$

15. $8p(-p^2 - 3p + 6)$

16. A rectangular deck measures $4x$ on one side and $3x + 6$ along the other. What is the area of the deck?

$$3x + 6 \quad \boxed{} \quad 4x$$

17. A rug is three times as long as it is wide. What is its area?

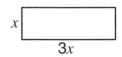

18. What property is used to multiply $x(3x + 2)$? _____

Multiplying Binomials

In arithmetic you have learned how to multiply and build a model for the problem 12×13.

$$
\begin{array}{rl}
12 & \rightarrow \quad 10 + 2 \\
\times\, 13 & \rightarrow \quad \underline{10 + 3} \\
& \quad\quad\; 6 \quad (3 \times 2) \\
& \quad\; 30 \quad (3 \times 10) \\
& \quad\; 20 \quad (10 \times 2) \\
& \underline{\; 100} \quad (10 \times 10) \\
& \quad 156
\end{array}
$$

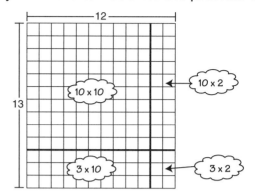

In algebra you can build a model and multiply for a similar problem $(x + 2)(x + 3)$.

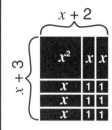

How many x^2 tiles? ___1___

How many x tiles? ___5___

How many units? ___6___

$\underline{1}\, x^2 + \underline{5}\, x + 6$

$$
\begin{array}{rl}
& x + 2 \\
\times & x + 3 \\
\hline
& \quad 6 \quad (3 \times 2) \\
& 3x \quad (3 \bullet x) \\
& 2x \quad (x \bullet 2) \\
\underline{x^2} & \qquad (x \bullet x) \\
\hline
x^2 + & 5x + 6
\end{array}
$$

Use tiles to model each problem. Draw a picture of the solution.

1. $(x + 3)(x + 1)$

x

How many x^2 tiles? _____

How many x tiles? _____

How many units? _____

$(x + 3)(x + 1) = $ _____ $x^2 +$ _____ $x +$ _____

2. $(x + 1)(x + 2)$

x

$(x + 1)(x + 2) = $ _____ $x^2 +$ _____ $x +$ _____

3. What is the pattern for multiplying two binomials? _____

_____

Multiplying Binomials

The model for multiplication is a rectangle.

You can build and model the multiplication of $(x + 3)(x - 1)$ with tiles.
Build $(x + 3)$ on one side of the rectangle and $(x - 1)$ on the other side of the rectangle.

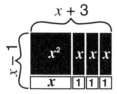

How many x^2 tiles? _____ x^2

How many x tiles? _____ x

How many units? _____

$(x + 3)(x - 1) =$ _____ $x^2 +$ _____ $x +$ _____

Multiplication uses the distributive property. Each term of the first binomial is multiplied by each term of the second binomial.

$$(x + 3)(x - 1) = x(x - 1) + 3(x - 1)$$
$$= x^2 - x + 3x - 3$$
$$= x^2 + 2x - 3$$

Use tiles to model each problem. Draw a picture of the solution. Shade positive models black and leave negative cubes white.

1. $(x + 1)(x - 2)$

   How many x^2 tiles? _____ x^2

   How many x tiles? _____ x

   How many units? _____

   $(x + 1)(x - 2) =$ _____ $x^2 +$ _____ $x +$ _____

2. Find the area of a rectangle whose length is $x + 2$ and width is $x - 1$.
   Draw a picture. Solve.

Multiply each term of the first binomial by each term of the second binomial. Combine like terms.

3. $(x - 3)(x + 1)$

4. $(2x + 2)(x + 1)$

Multiplying Binomials Using **FOIL**

To multiply two binomials, multiply each term of the first term by each term of the second term. Then combine like or similar terms.

<table>
<tr><td>The Long Method</td><td>The Short Method: Using FOIL</td></tr>
</table>

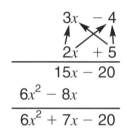

$$\begin{array}{r} 3x \quad - 4 \\ 2x \quad + 5 \\ \hline 15x - 20 \\ 6x^2 - 8x \\ \hline 6x^2 + 7x - 20 \end{array}$$

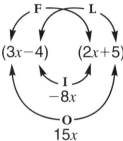

$(3x-4)$ $(2x+5)$

$-8x$

$15x$

First terms: $3x \cdot 2x = 6x^2$

Outer and Inner terms: $-8x + 15x = 7x$

Last terms: -20

sum: $6x^2 + 7x - 20$

Find each product.

1. $(x + 1)(x - 3)$

2. $(m - 2)(m - 5)$

3. $(x - 3)(2x + 3)$

4. $(3y - 6)(4y + 2)$

5. $x(3x - 6)$

6. $(6a - 3)(6a - 3)$

7. $(2x - 3)(2x + 3)$

8. $(-4y + 2)(4y + 2)$

9. $(6b + 5)(5b - 2)$

10. $(7x - 4)(x - 7)$

11. Which is the correct product of $(x + 3)(x - 2)$?

    a. $x^2 - 6$ b. $x^2 + x + 6$ c. $x^2 + x - 6$ d. $x^2 + 5x - 6$

12. A field measures $(2x + 10)$ by $(3x + 6)$. What is the area?

    a. $6x^2 + 42x + 60$ b. $6x^2 + 16$ c. $6x^2 + 60$ d. $6x^2 + 5x + 16$

Squaring Binomials

You know that $(x+2)^2$ means $(x+2)(x+2)$.

Model the multiplication.

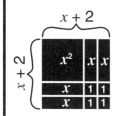

$x + 2$

____1____ How many x^2?

____4____ How many x?

____4____ How many units?

$(x + 2)^2 = x^2 + 4x + 4$

To square a binomial:

1. Square the first term. $\quad\quad\quad\quad\quad\quad (x^2)$

2. Double the product of the two terms. $(2 \bullet 2x)$

3. Add the square of the last term. $\quad\quad (2^2)$

$(x + 2)^2 = x^2 + 4x + 4$

Make a model to help solve. Draw a picture of the solution.

1. $(x + 3)^2$ $\qquad\qquad\qquad\qquad$ $(x + 3)^2 =$ _____

Use the three steps shown at the top of the page to square each binomial.

2. $(x - 2)^2 =$ ___ + ___ + ___ $\qquad$ **3.** $(a + 3)^2 =$ $\qquad\qquad\qquad$ **4.** $(m - 5)^2 =$

5. $(2x - 3)^2 =$ $\qquad\qquad\qquad$ **6.** $(x - 6)^2 =$ $\qquad\qquad\qquad\qquad$ **7.** $(5 - c)^2 =$

8. $(x + y)^2 =$ $\qquad\qquad\qquad\quad$ **9.** $(x - y)^2 =$ $\qquad\qquad\qquad\qquad$ **10.** $(4x - 1)^2 =$

12. In general, $(a + b)^2 = a^2 + 2ab + b^2$. Complete this picture of the square of $a + b$ to show the product.

$a + b$

$a \quad\quad b$

$a + b$

a

b

Algebraic Fractions

The most important principle of fractions used in arithmetic and algebra is: the value of a fraction is not changed when you multiply or divide both terms by the same number (except 0). In fact, you are really multiplying or dividing the fraction by 1.

Example: $\dfrac{2}{3} \underset{\times 2}{\overset{\times 2}{}} = \dfrac{4}{6}$ and $\dfrac{4}{6} \underset{\div 2}{\overset{\div 2}{}} = \dfrac{2}{3}$

Check by using cross products:

$\dfrac{2}{3} \times \dfrac{4}{6}$

$2 \cdot 6 = 3 \cdot 4$

$12 = 12$

Multiply the numerator of one fraction times the denominator of the other and vice versa.

Example: Simplify. $\dfrac{10}{x} = \dfrac{5}{x-2}$

Method 1: Multiply both sides by the same term.

$\dfrac{(10)}{(x)} = \dfrac{(5)}{(x-2)}$

Multiply each side by both denominators.

$\dfrac{\cancel{(x)}}{1} \cdot \dfrac{(x-2)}{1} \cdot \dfrac{(10)}{\cancel{(x)}} = \dfrac{(5)}{\cancel{(x-2)}} \cdot \dfrac{\cancel{(x-2)}}{1} \cdot \dfrac{(x)}{1}$

Cancel the denominators.

$10\,(x-2) = 5x$

$10x - 20 = 5x$

$x = 4$

Solve for x.

Method 2: Cross products.

$\dfrac{(10)}{(x)} \times \dfrac{(5)}{(x-2)}$

$5x = 10\,(x-2)$

$5x = 10x - 20$

$5x = 20$

$x = 4$

Simplify each equation by eliminating denominators.

1. $\dfrac{2}{x} = \dfrac{4}{x+5}$

2. $\dfrac{12}{y} = \dfrac{6}{y-3}$

3. $\dfrac{2}{a} = \dfrac{5}{a+3}$

4. $\dfrac{6}{n} = \dfrac{3}{n-2}$

Choose the equation equivalent to the given equation.

5. $\dfrac{17}{x} = \dfrac{5}{x-1}$

    A $5x = 17x - 1$
    B $5x - 5 = 17x$
    C $5x = 17x - 17$
    D $5x = x - 17$

_____

6. $\dfrac{m}{m-1} = \dfrac{9}{17}$

    A $17 = 9m - 9$
    B $17m = 9m - 9$
    C $17m - 17 = 9m$
    D $17m - 1 = 9m$

_____

7. $\dfrac{6}{n+3} = \dfrac{12}{n+4}$

    A $12n + 48 = 6n + 18$
    B $12n + 36 = n + 4$
    C $12n + 3 = 6n + 4$
    D $12n + 36 = 6n + 24$

_____

Equations Involving Absolute Value

The **absolute value** of a number is its distance from zero on a number line.
The symbol for absolute value is two vertical bars around the number: |4|.
Equations involving variables and absolute values will have two numbers as correct answers.

If $x = |4|$, x may be $+4$ or x may be -4. The solution is shown on the graph.

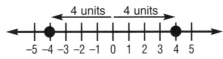

$|x - 2| = 1$ also means $x - 2 = 1$ or $-(x - 2) = 1$.

Solve the positive case: $x - 2 = (+1)$ Solve the negative case: $-x + 2 = 1$
$$x = 3$$
$$1 = x$$

The solution set is {3,1}.

Solve each problem and graph the solutions.

1. $|a + 2| = 9$ **2.** $|x| = 0$ **3.** $|y - 7| = 2$

4. $|m - 10| = 7$ **5.** $|n + 5| = 5$ **6.** $|8 - t| = 3$

7. $12 = |x + 4|$ **8.** $|3 + b| = 7$ **9.** $5 = |x + 7|$

Match the graph with the equation.

10. $|x - 2| = 4$

a.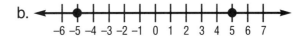

11. $|2 - x| = 3$

b.

12. $|x| = 5$

c.

13. $|x + 3| = 1$

d.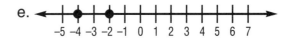

14. $|x + 2| = 3$

e.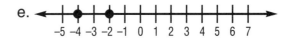

Two-Step Equations Involving Absolute Value

The **absolute value** of a number is its distance from zero on a number line.
If $|x| = 3$, x may equal $+ 3$ or $- 3$. $x = \pm 3$.

Equations having variables and absolute values have two correct answers.
Solve the equation as usual:

If $3|x| + 5 = 8$
    $3|x| + 5 - 5 = 8 - 5$
    $$\frac{3|x|}{3} = \frac{3}{3}$$

Subtract 5 from both sides of the equation. Divide both sides by 3.

    $|x| = 1$ $x = 1$ or $x = -1$

Solve each problem. Graph the solution.

1. $4 + 7|m| = 18$

2. $|g| + 3 = 6$

3. $4|x + 3| = 12$

4. $|y| + 5 = 6$

5. $36 = 4|s| + 20$

6. $9 + 5|x| = 29$

7. $15 = 3|b|$

8. $|m| - 10 = 2$

9. $8 + 6|z| = 20$

10. $- 4 + |2x + 4| = 14$
HINT: $(2x + 4) = 18$ or $-(2x + 4) = 18$

11. $|4r - 8| - 3 = 9$

12. $9 = |14 + 2y| + 7$

Choose the replacement set that makes the statement true.

13. $|x - 6| = 3$ **A** $\{-3, 9\}$ **B** $\{3, 9\}$ **C** $\{-3, 2\}$ **D** $\{0, 6\}$

14. $2|r + 3| - 5 = 7$ **A** $\{3, -9\}$ **B** $\{-3, -9\}$ **C** $\{-3, 2\}$ **D** $\{3, 2\}$

Inequalities Involving Absolute Value

Inequalities involving absolute value may be represented on a number line.

$|x| < 3$ means the distance from 0 to x is less than 3 units.

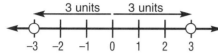

$$-3 < x < 3$$

Solve $|2x-4| + 1 \geq 9$

> Subtract 1 from both sides.

$$|2x-4| \geq 8$$

> Set up a positive case and a negative case.

Positive $2x - 4 \geq 8$ Negative $-(2x-4) \geq 8$
$$2x \geq 12$$ $$-2x+4 \geq 8$$
$$x \geq 6$$ $$-2x \geq 4$$
$$x \leq -2$$

> Reverse the inequality sign when dividing by a negative number.

Solve and graph the solution.

1. $2|y| \leq 10$

2. $|2b| < 10$

3. $|x - 2| = 5$

4. $|3x| + 2 > 3$

5. $|x + 7| > 4$

Direct Variation

The distance *d* traveled at a fixed rate *r* varies directly with the time *t* spent travelling. The relationship between the distance traveled and the time spent is shown by the equation $d = rt$.

A car travels at a constant rate of 60 miles per hour. How far will the car travel in 2 hours? 3 hours? 4 hours? The table shows the number of miles traveled.

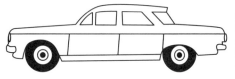

t (hours)	1	2	3	4
d (distance)	60	120	180	240

Words like *rate*, *average speed* and converting measurements are examples of direct variation.

Use direct variation to solve the following questions.

1. Water fills a 60-gallon tub at the rate of 2.5 gallons per minute. How long will it take to fill the tub at this rate?

2. There are 12 inches in 1 foot. How many feet are there in 150 inches?

3. On vacation, the Carters drove 618 miles the first day. If they drove for 12 hours, what was their average speed?

4. Olivia bicycled 22 minutes at a rate of .25 miles per minute. How far did she ride?

5. At the beauty salon, shampoo is sold in 1-cup bottles. Yesterday, the salon bottled 900 oz. of shampoo. If 1 cup = 8 oz., how many bottles of shampoo were filled yesterday?

6. A quart holds 32 oz. How many quart containers will be needed to hold 352 ounces of orange juice?

7. If Del types 53 words per minute, how long will it take him to type his 780-word report?

8. A car travels a distance of 275 miles in $4\frac{1}{2}$ hours. What is the average speed per hour?

Direct Variation and Proportion

In **direct variation**, each quantity changes in proportion to the other. When one quantity increases (or decreases), the other quantity increases (or decreases).

Bryan's wages vary directly with the number of hours that he works. If his wages w for 8 hours h are $104.00, how much will he earn in 40 hours?

Let w = wages. Set up a proportion comparing wages to hours.

$$\frac{\text{wages}}{\text{hours}} \qquad \frac{104.00}{8} = \frac{w}{40}$$

Use cross products to solve:

$$\frac{104.00}{8} \diagdown\diagup \frac{w}{40}$$

$$8w = 40 \cdot 104$$
$$8w = 4160$$
$$w = \$520$$

Set up a proportion. Solve.

1. A cruise ship consumes 54 gallons of fuel in 3 hours. At that same rate, how many gallons will be consumed in 4.5 hours?

2. On the first day of her driving trip, Olivia travelled 720 miles in 12 hours. At that same rate, how long did it take her to drive 540 miles the following day?

3. The ratio of oil to vinegar in a certain brand of salad dressing is 7:5. Using this ratio, how much oil must be blended with 15 quarts of vinegar?

4. Bridgette worked 30 hours and earned $240. How much would Bridgette make if she worked 25 hours?

5. PlutoVision manufactures televisions. They can make 200 televisions in a 40-hour work week. If the hours are cut to 35 per week and work is done at the same rate, how many televisions will be made each week?

6. The cook at a summer camp ordered enough food to feed 300 campers for 14 days. If food for 350 campers was delivered, how many days would the food last?

Direct Variation: Charts, Tables, Equations

Direct variation means that as one factor increases, the other factor also increases. We represent direct variation by an equation in the form $y = kx$, where k is <u>not</u> zero. k is called the **constant of variation.**

Marti's aunt runs a small hardware store. Marti sorts bolts of various sizes into bins labeled with the bolt size and price per bolt. Marti enters the price per bolt into the cash register and then the number of bolts purchased to find the total price. Marti saw the total cost was a relationship between the price per item and the number of items purchased.

Using t to represent the total price, p for the price per item and n for the number purchased, she wrote these algebraic sentences: $t = pn$ or $p = \dfrac{t}{n}$

Ounces Plant Food	Ounces Water
1	4
2	8
3	12
4	16
5	20
6	24
7	28
8	32
9	36
10	40
11	
12	
13	
14	
15	
16	

The plant food sold in the store is concentrated and must be mixed with water according to a chart before it is sprayed on the plants. When Marti sold the last bag of Great Green Plant Food to a customer, she noticed the bottom of the chart had been torn off, so she helped the customer figure out the missing numbers.

1. What were the missing numbers? How do you know?

_____

_____

_____

2. Which statement about the amount of plant food and the amount of water best describes the pattern in the chart?

_____

A The more plant food used, the less water used.

B The more plant food used, the more water used.

C The amount of water remains the same as the amount of plant food increases.

D The amount of water increases as the amount of plant food decreases.

3. Use p to represent ounces of plant food and w to represent ounces of water. Write an algebraic equation showing how p and w are related.

_____

4. What is the constant of variation for the equation in problem 3?

_____

Direct Variation: Graphs

Graphs and charts are two ways of showing relationships. Algebraic statements are another. Using w to represent ounces of water and p to represent ounces of plant food, an algebraic statement can be written to show the relationship between water and plant food in the mixture as $w = 4p$.

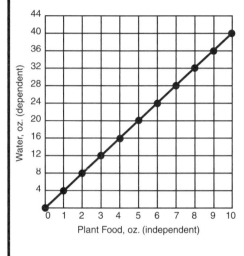

The amount of plant food is the **independent variable.** The amount of water is the **dependent variable,** its value depends upon how much plant food is used.

Some containers of plant food in the store had a different kind of chart. Marti recognized it as a **graph.** Compare the graph to the chart on the previous page. How does the information on the graph relate to the information on the chart?

_____

One oz. of plant food needs _____ oz. of water.

Six oz. of plant food needs _____ oz. of water.

Make a chart to show the relationship. Write an algebraic statement of the relationship. Graph and list the ordered pairs in your graph.

1. One pint has 2 cups.

pt.	cup	$p = 2c$
1	2	(1, 2)
2		(2, 4)
3		(3, 6)

relationship: _**p = 2c**_

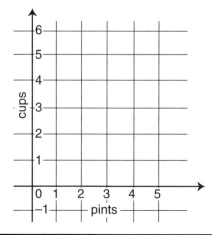

2. Pete's hourly wage is $8.00.

hr.	wage	_____
1		(___, ___)
2		(___, ___)
3		(___, ___)

relationship: _____

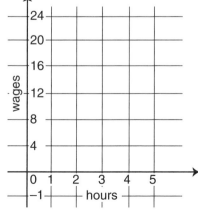

3. The circumference of a circle is equal to about 3.14 times the diameter.

dia.	circum.	_____
1		(___, ___)
2		(___, ___)
3		(___, ___)

relationship: _____

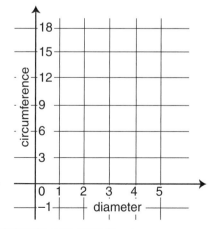

Graphing Direct Variation

A **direct variation** is described by an equation of the form $y = kx$, where $k \neq 0$. k is called a **constant of variation.** To find the constant of variation, divide each side by x: $\dfrac{y}{x} = k$

The library is holding a used book sale. All books are being sold for $2.00 each. If t = total cost and b = number of books, write an equation to show the relationship.

Make a table and draw a graph to show the cost of 1 to 5 books.

b	t
1	**2**
2	
3	
4	
5	

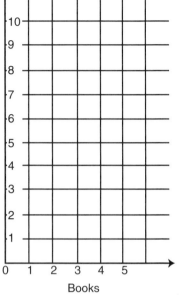

For each statement, write an equation, make a table and graph 3 sets of ordered pairs.

1. The sum of 2 numbers is 4.

x	y
1	
2	
3	

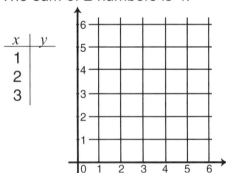

2. There are 3 feet in 1 yard.

$yd.$	$ft.$
1	
2	
3	

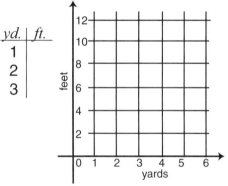

3. The perimeter of a square is equal to four times the length of one side.

s	P
1	
2	
3	

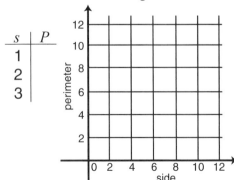

4. There are 2 cups in 1 pint.

p	c
1	
2	
3	

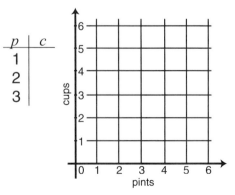

5. What shape is the graph of a direct variation? _____

Modeling a Situation

Graphs tell stories. Graphs, with or without scales on the axis, model a situation to help us visualize what is happening.

Lenore begins her morning run by steadily increasing her rate of speed, then levels off and runs at a steady speed. The graphic model looks like this:

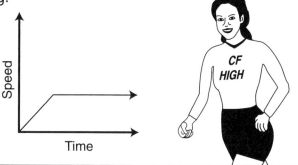

1. Which of the following graphs shows Lenore running up hill at a steady pace, then speeding up as she runs down the hill? _____

A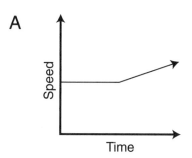

B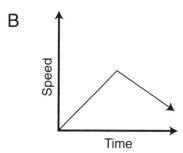

2. Give a possible explanation of what is happening in the other graph.

   _____

   _____

3. Which of the following best describes what is happening in the graph below? _____

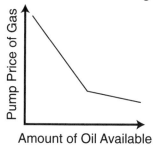

   A When oil is plentiful, gas prices are higher.

   B When oil is plentiful, gas prices are lower.

   C When oil is plentiful, prices do not change.

   D When oil is plentiful, gas companies advertise more.

4. Draw a graph showing the elevation of Chuck's bike when he rides up a steep hill for 5 minutes, bikes on a flat road for 5 minutes, then coasts down a gradual hill for 10 minutes.

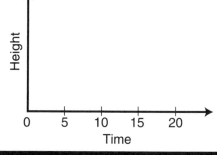

Mixture Problems

Mixtures are made when different ingredients are combined to make a blend. Drawing pictures helps set up the equation.

A snack mix is to be made with almonds and peanuts. Peanuts cost $3.00 per pound and almonds cost $7.00 per pound. A mixture of 100 pounds costing $6.00 per pound is to be made. How many pounds of peanuts and almonds are needed?

$$\boxed{\begin{array}{c} p \\ @\ \$3 \end{array}}_{\text{peanuts}} + \boxed{\begin{array}{c} 100-p \\ @\ \$7 \end{array}}_{\text{almonds}} = \boxed{\begin{array}{c} 100 \\ @\ \$6 \end{array}}_{\text{mixture}}$$

$$3 \cdot p + 7(100 - p) = 6 \cdot 100$$
$$3p + 700 - 7p = 600$$
$$\frac{-4p}{-4} = \frac{-100}{-4}$$

Let p = pounds of peanuts

$100 - p$ = pounds of almonds

p = 25 pounds of peanuts

$100 - p$ = 75 pounds of almonds

Draw a picture. Write an equation to solve each problem.

1. The coffee shop makes several blends. One blend uses a special Hawaiian coffee, which costs $9.00 a pound and a Columbian coffee, which costs $5.00 per pound. The cost of the mixture is $6 per pound. How much of each coffee would it take to make 400 pounds of mixture?

2. A cheese shop makes pizza topping for the pizza shop next door. The pizza shop needs 20 pounds of cheese at a cost of $5 per pound. They combine mozzarella cheese at $3 per pound with a provolone cheese at $7 per pound. How much of each do they use?

3. A caterer makes a fruit punch mix, which costs $3.00 per gallon from cranberry juice costing $2.50 per gallon and orange juice costing $4.00 per gallon. The caterer needs 250 gallons of punch. What amounts of the two juices will be used?

4. Twenty pounds of meatloaf will be made at a cost of $3.50 per pound. Ground pork selling for $3.00 per pound will be mixed with ground beef costing $3.80 per pound. How much of each meat will be used?

Work Problems

Work problems deal with the amount of time needed to do a piece of work.

Sara can mow the yard in 4 hours and Seth can mow the same yard in 6 hours. How long will it take to mow the lawn if they work together?

Let x = length of time needed to finish the work when they work together.

$\frac{1}{4}x$ = part of the yard done by Sara in x hours

$\frac{1}{6}x$ = part of the yard done by Seth in x hours

$$\frac{1}{4}x + \frac{1}{6}x = 1$$ Multiply both sides by 12 to get rid of the denominators.

$$12 \cdot \frac{1}{4}x + 12 \cdot \frac{1}{6}x = 12 \cdot 1$$

$$3x + 2x = 12$$ Add like terms.

$$5x = 12$$ Divide both sides by 5.

$$x = 2\frac{2}{5} \text{ hr.}$$

Solve these problems following the steps in the example.

1. Beth can paint the living room in 5 hours. It takes Liz 8 hours to paint the room. How quickly can they get the room painted if they work together?
   Let x = time needed together

2. Marti is making costumes for the school play. He knows it will take him 10 hours to make the costumes. Tim can make the costumes in 12 hours. How long will it take if they work together?

3. Dana can rake the lawn in 4 hours and Emily can rake it in 3 hours. How much time will they spend raking if they work together?

4. Drew can plant the garden in 8 hours. Rob can plant it in 10 hours. Working together, how long will it take them to plant the garden?

Motion Problems

The formula for finding distance traveled when the rate and time of travel are known is $d = rt$ (distance = rate × time). Problems of distance, rate and time are called **motion problems.**

Draw a picture to help visualize the problem and set up the equation.

Cyclist A leaves town riding 21 miles per hour. Four hours later, cyclist B leaves from the same starting point traveling 35 miles per hour. How long will it take B to catch up with A?

Let t = A's time.
Let $t - 4$ = B's time.

	Rate	Time	Distance
A	21	t	$21t$
B	35	$t - 4$	$35(t - 4)$

Cyclist A's distance = cyclist B's distance.

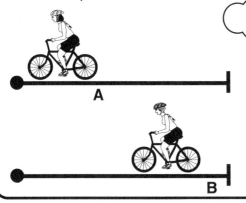

A

B

$21 \cdot t = 35(t - 4)$
$21t = 35t - 140$
$-14t = -140$
$t = 10$ hr.
A's time is 10 hours.

$t - 4 = 6$ hr.
B's time is 6 hours.

Check your answer:
$21 \cdot 10 = 35(10 - 4)$
$210 = 35 \cdot 6$
$210 = 210$

Draw a Picture. Solve the problems. Check your answers.

1. Monica left on the bus traveling 45 mph. Two hours later Joey left from the same location, traveling the same roads to catch up to her. He drives 60 mph. How long will it take him to catch up with Monica?

2. Kyla leaves on a bike riding at 10 mph. Three hours later, Chip leaves from the same place driving 50 mph. How long will it be before Chip overtakes Kyla? How far from their departure point will they be when they meet?

3. Train A leaves headed west at 33 mph. Two hours later, express train B leaves from the same station on a parallel track traveling west at 66 mph. How long until B overtakes A?

4. A cyclist averages 12 mph. Four hours later, a motorcyclist leaves from the same spot traveling an average of 50 mph. At what distance from the starting point will the motorcyclist pass the cyclist?

Motion in an Opposite Direction

Some motion problems involve two things moving in opposite directions from the same point. The formula for finding distance traveled when the rate and time of travel are known is $d = rt$ (distance = rate × time).

Sometimes we know the distance traveled and need to find the speed traveled.

Plane A left the airport at the same time as plane B going in opposite directions. Plane B was traveling 150 mph faster than plane A. In 3 hours they were 4,500 miles apart. What was the average speed of each plane?

Let x = rate of airplane A
$x + 150$ = rate of airplane B

Rate	× Time	= Distance
A	x	3	$3x$
B	$x + 150$	3	$3(x + 150)$

$$3 \bullet x + 3(x + 150) = 4500$$
$$3x + 3x + 450 = 4500$$
$$6x = 4050$$
$$x = 675 \text{ mph for plane A}$$
$$x + 150 = 825 \text{ mph for plane B}$$

Check your answer:
$$3 \bullet 675 + 3(675 + 150) = 4500$$
$$2025 + 2475 = 4500$$
$$4500 = 4500$$

Solve these problems by following the steps in the example. Check your answers.

1. Two trains left the station going in opposite directions. Four hours later they were 336 miles apart. Train A was traveling 16 mph faster than train B. Find the speed of each train.

2. A bike rider and a person on a scooter left the grocery store at the same time. The person on the scooter traveled 7 mph faster than the cyclist. In 2 hours they were 58 miles apart. What was the speed of each?

3. Starting from the same spot but traveling in opposite directions, two runners leave at the same time. Runner A is averaging 2 mph faster than runner B. After 1.5 hours they are 21 miles apart. Find the speed of each runner.

4. Truck A headed east from Phoenix at the same time as truck B headed west from the same gas station. Truck A averaged 11 mph slower than truck B. In 7 hours they were 931 miles apart. How fast was each truck going?

Points on a Line

The degree of an equation is the same as that of its highest degree term.

$$\text{first-degree equation: } x + 2y = 7$$
$$\text{second-degree equation: } x^2 + 4xy + 2y^2 = 0$$
$$\text{third-degree equation: } x^3 - x^2y + y^3 = 0$$

A first-degree equation is called a **linear equation** since its graph is always a straight line.

Example: Graph the equation $x + y = 6$. Make a table and plot 3 points that make the equation true.

x	y
0	6
2	4
4	2

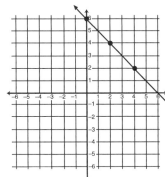

Does point (3, 3) lie on the line $x + y = 6$?

Does point (7, 1) lie on the line $x + y = 6$?

To check if a point lies on a line, substitute the value of the points into the equation.

Check:
$$x + y = 6$$
$$3 + 3 = 6$$
$$6 = 6$$
Yes.

Check:
$$x + y = 6$$
$$7 + 1 = 6$$
$$8 \neq 6$$
No.

For each equation, determine which points lie on the given line.

1. $2x + y = 2$
   A (3, 8)
   B (1, 4)
   C (−1, 0)
   D (0, 2)

   _____

2. $x + 2y = 4$
   A (0, 3)
   B (2, 1)
   C (7, −3)
   D (−8, 4)

   _____

3. $2x + 1y = 1$
   A (0, 4)
   B (1, −1)
   C (−1, 1)
   D (3, 0)

   _____

4. $y = x + 4$
   A (0, −4)
   B (−1, 2)
   C (0, 4)
   D (3, 5)

   _____

5. The point (3, 2) lies on which line?
   A $y = 2x - 4$
   B $y = 2x + 4$
   C $y = \frac{1}{3}x + 2$
   D $y = \frac{1}{3}x - 3$

   _____

6. The point (1, −3) lies on which line?
   A $y = x + 1$
   B $y = x - 2$
   C $y = x + 4$
   D $y = x - 4$

   _____

Relating Points on a Graph to Equations

If a true statement results when the numbers in an ordered pair are substituted into an equation with two variables, the ordered pair is a solution to the equation.

You can construct a graph from an equation.

Example: $y = x + 1$

x	y
0	1
1	2
−1	0

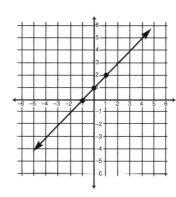

Does the point (3, 4) lie on the line?

By examination, (3, 4) does.

To check, substitute 3 and 4 for x and y in the original equation.

$$y = x + 1$$
$$4 = 3 + 1$$
$$4 = 4$$

Yes

Circle the points that lie on the equation of the line. Use those points to graph the line.

1. $-2x + y = 2$
(3, 8) (1, 4) (−1, 0)

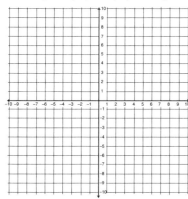

2. $x + 2y = 4$
(0, 2) (−4, 4) (−8, 2)

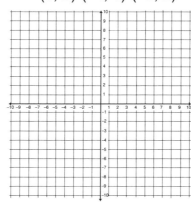

3. $2x + y = 1$
(1, −1) (−1, 1) (2, −3)

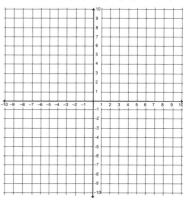

4. $-5x - 3y = 2$
$(0, -\frac{2}{3})$ $(-\frac{1}{5}, -\frac{1}{3})$ (2, −4)

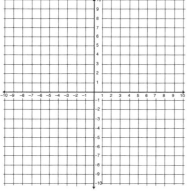

5. $y = x + 4$
(0, 4) (−1, 2) (−4, 0)

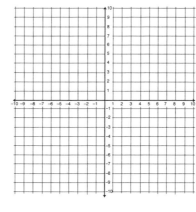

6. $\frac{y}{2} = x - 3$
(0, 3) (3, 0) (−1, −8)

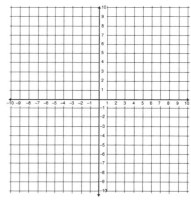

Intercepts of a Linear Equation

A **linear equation** is an equation whose graph is a line. Linear equations may contain one or two variables with no variable having an exponent greater than one.

$y = 2x + 4$ is a **linear equation** written in slope-intercept form.

One method of graphing a linear function is to find its x- and y-**intercepts** and a third point to check the line.

The **y-intercept** is the y-coordinate of the point where a line crosses the y-axis.

Think: when $x = 0$, what does y equal?

$y = 2x + 4$
$y = 0 + 4$
$y = 4$

x	y
0	4
−2	0
1	6

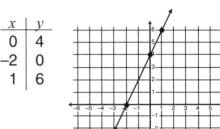

The **x-intercept** is the x-coordinate of the point where a line crosses the x-axis.

Think: when $y = 0$, what does x equal?

$y = 2x + 4$
$0 = 2x + 4$
$2x = -4$
$x = -2$

For each equation, make a table with 3 points. Find the x- and y-intercepts and a third point for each line. Plot the points and draw the line.

1. $x + y = 5$

x	y
0	—
—	0
—	—

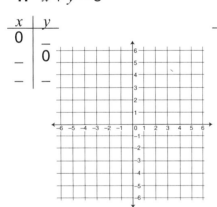

2. $y = x + 2$

x	y
0	—
—	0
—	—

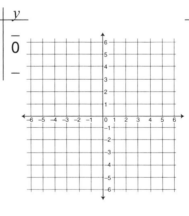

3. $y = x - 2$

x	y
0	—
—	0
—	—

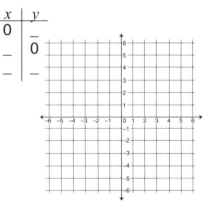

4. $y = 3x$

x	y
0	—
—	0
—	—

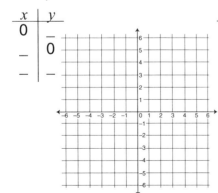

5. $y = 2x$

x	y
0	—
—	0
—	—

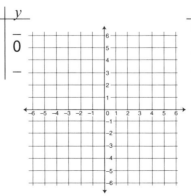

6. $y = 2x - 1$

x	y
0	—
—	0
—	—

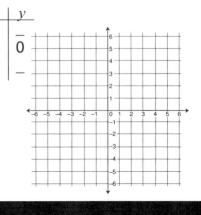

Linear Equations Parallel to x or y Axis

Some linear equations have only one variable. No matter what you choose for the missing variable, the other axis has a constant value.

Graph the equation $x = -2$.

If $x = -2$, no matter what you choose for y, x has the value of -2.

A chart might be:

x	y
-2	0
-2	1
-2	2

We only need the first 3 points to be sure of our line.

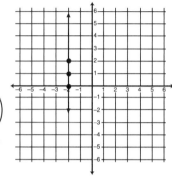

The graph of $x = -2$ is a vertical line and is parallel to the y axis.

Graph the equation $y = 5$.

Complete the chart for $y = 5$:

x	y
0	5
1	5
2	5

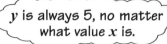

y is always 5, no matter what value x is.

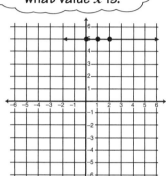

The graph of $y = 5$ is horizontal and lies parallel to the x axis.

For each equation, make a table of 3 points. Graph the points and draw the line formed. Tell whether the line is horizontal, vertical or neither.

1. $x = 3$ __vertical__

x	y
3	
3	
3	

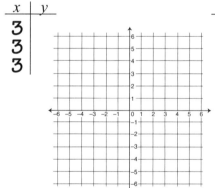

2. $y = -1$ _____

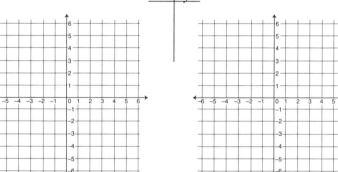

3. $x = -1$ _____

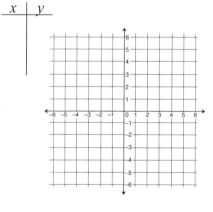

4. $y + 1 = 4$ _____

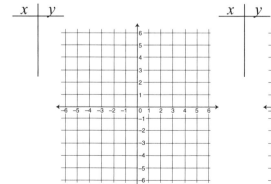

5. $x = -3$ _____

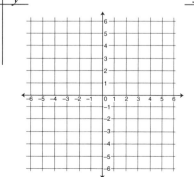

6. $4x - 8 = 0$ _____

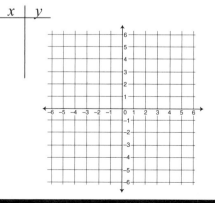

Model of Slope of a Line

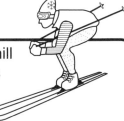

When you think of slope, you might think of a skier racing downhill or a biker going up a steep hill. Most graphs slant up or down as you look from left to right. In math we can model slope on a coordinate grid.

We can use a geoboard to model a line and its slope. The lower left corner of the geoboard is (0,0). Label the 5 points in a horizontal and vertical direction as shown. Find the slope of line AB between point A (1,1) and point B (3,4).

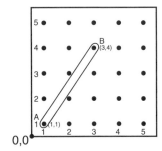 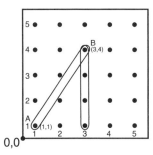 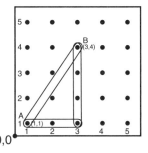

Form a right triangle to find the vertical change and the horizontal change. As we move from point A to point B, point B is 3 units higher (rise) and 2 units to the right (run). This change is called the **slope** of the line.

> ***Slope*** of a line is the ratio of the $\dfrac{\text{rise}}{\text{run}}$ or $\dfrac{\text{change in } y}{\text{change in } x}$ or $\dfrac{y_2 - y_1}{x_2 - x_1}$
>
> The slope of the line we constructed is $\dfrac{4-1}{3-1} = \dfrac{3}{2}$

Use the geoboard to find the slope of a segment with the two given endpoints. Draw the segment and compute the slopes.

A (1,2) B (3,5)   A (5,2) B (1,4)   A (2,3) B (5,3)   A (1,4) B (2,3)

1. 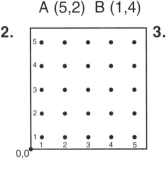   2.   3.   4.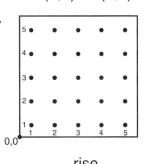

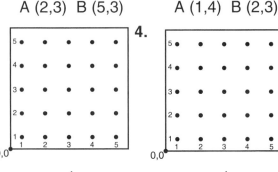

slope = $\dfrac{\text{rise}}{\text{run}}$ =   slope = $\dfrac{\text{rise}}{\text{run}}$ =   slope = $\dfrac{\text{rise}}{\text{run}}$ =   slope = $\dfrac{\text{rise}}{\text{run}}$ =

Identifying Slope from a Graph

To find the slope of a line from a graph, make a right triangle using two points on the line. As we "run" from left to right along the x-axis 1 unit, how much do we "rise" (or fall) along the y-axis? What is the slope of each line?

A

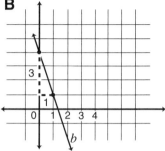

Slope of line *a*

$$\frac{\text{rise}}{\text{run}} = \underline{\hspace{1cm}}$$

B

Slope of line *b*

$$\frac{\text{rise}}{\text{run}} = \underline{\hspace{1cm}}$$

C

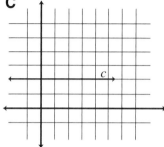

Slope of line *c*

$$\frac{\text{rise}}{\text{run}} = \underline{\hspace{1cm}}$$

Conclusions: Fill in the letter of the correct graph.

Graph _____ slopes *up* from left to right so it has a **positive** slope.

Graph _____ slopes *down* from left to right so it has a **negative** slope.

Graph _____ is horizontal and *parallel to* the x-axis, so it has a **zero** slope.

Find the slope of each line. Slope = $\dfrac{y_2 - y_1}{x_2 - x_1}$

1.

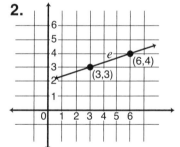

Positive or negative? _____

Slope of *d* = _____

2.

Positive or negative? _____

Slope of *e* = _____

3.

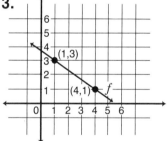

Positive or negative? _____

Slope of *f* = _____

4. What is the slope of line *g*?

_____

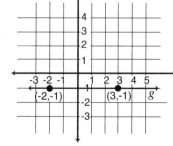

5. What is the slope of line *h*?

_____

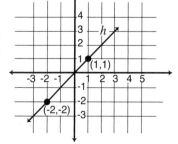

Slope of a Line

Sara and Devon decided to make a graph showing the relationship between gallons and quarts.

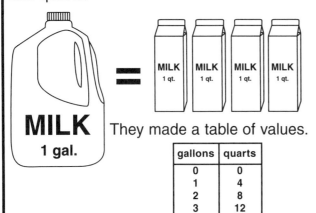

They made a table of values.

gallons	quarts
0	0
1	4
2	8
3	12

Sara said, "We can write an algebraic equation. If we let x be 'gallons', and let y be 'quarts', my equation is $y = 4x$."

Devon said they could graph the points to find the slope. From the table they wrote four sets of *ordered pairs*.

Complete the sets: (0,0) (1,4) (2,8) (_____, _____)

Graph the line.

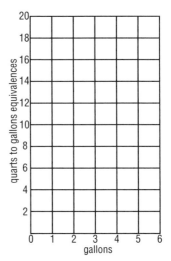

To compute the slope, they selected two points on the graph, (1,4) and (2,8).

$$\text{Slope} = \frac{\text{change in } y}{\text{change in } x} = \frac{8 - 4}{2 - 1} = \frac{4}{1} = 4$$

They saw the slope of 4 was related to the equation $y = 4x$.

Graph the points. Draw the line. Find the slope of each line.

1. (2,1) (4,4)

slope = _____

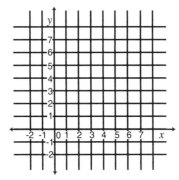

2. (1,4) (3,4)

slope = _____

3. Find the slope of a line comparing ounces to pounds.

slope = _____

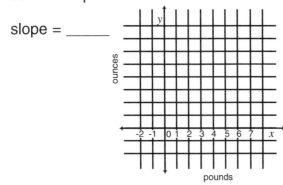

4. Find the slope of a line comparing yards to feet.

slope = _____

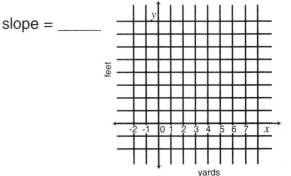

Slope-Intercept Form of Linear Equations

Linear equations contain one or two variables, with no variable having an exponent greater than one. The standard form for a linear equation is $4x + 2y = 6$.

An equation may also be written in slope-intercept form by solving for y:

$y = \mathbf{m}x + \mathbf{b}$ $\mathbf{m}$ tells us the slope and $\mathbf{b}$ tells us the y-intercept

To put $2y - 2x = 8$ in slope-intercept form, y must be by itself on one side of the equation.

$2y - 2x + 2x = 2x + 8$ ← Add $2x$ to each side.

$2y = 2x + 8$

$\dfrac{2y}{2} = \dfrac{2x}{2} + \dfrac{8}{2}$ ← Divide by 2.

slope $\underset{\nearrow}{\quad}$ $y = 1x + 4$ $\underset{\nwarrow}{\quad}$ y-intercept

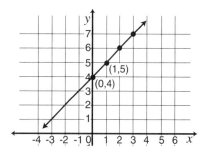

We can use the slope and y-intercept to graph the line. First, plot (0,4). The slope is 1 so rise/run = 1 which means we go up 1 unit and across 1 unit to plot another point on the line. We can then draw the line connecting them.

To check our solution, pick another point on the line to see if it satisfies the equation. Does (2, 6) satisfy the equation?
Check:

Write each equation in slope-intercept form. Use the slope and y intercept to graph the line. Pick another point on the line you graphed and check to see if it also satisfies the equation.

1. $y - 4x = 2$

$y =$ _____

y-intercept = _____

slope = _____

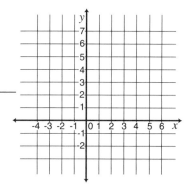

2. $2y + 4x = 4$

$y =$ _____

y-intercept = _____

slope = _____

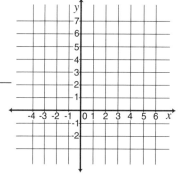

3. $y - x = 4$

$y =$ _____

y-intercept = _____

slope = _____

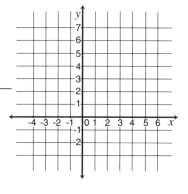

4. $y - x = 5$

$y =$ _____

y-intercept = _____

slope = _____

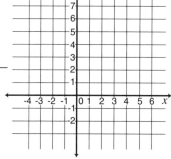

Graphing a Line, Given the Slope and a Point on the Line

Graph the line passing through (1,1) with a slope of 3. Write the equation of the line. Find the y-intercept for the line.

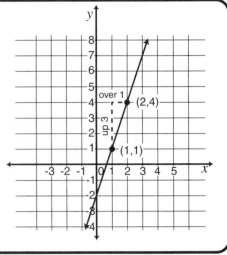

- Start at point (1,1). Plot and label the point.
- The slope of 3 means it rises 3 for each run of 1.
- Find a second point that is 3 points higher and 1 point to the right. This gives a second point on the line and allows the line to be drawn.

Slope = _____

Equation of line _____

Given a point and the slope m for a line, draw the graph formed. Write the equation for the line.

1. (3, 3) $m = 1$

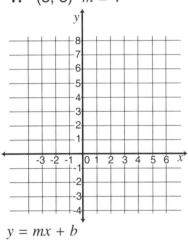

$y = mx + b$

$y = $ _____

2. (3, 2) $m = 2$

$y = mx + b$

$y = $ _____

3. (−2, 5) $m = −1$

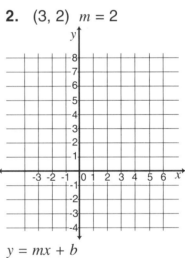

$y = mx + b$

$y = $ _____

4. (0, 0) $m = \dfrac{2}{5}$

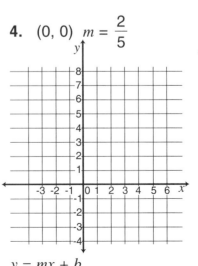

$y = mx + b$

$y = $ _____

Slopes of Parallel Lines

Parallel lines are lines that lie in the same plane but never intersect. The sides of a ladder are real-world examples of the idea of parallel lines.

Three equations related to lines a, b and c are written in a special form called the slope-intercept form. The graph of each line is shown on the grid.

slope $\underbrace{y = \mathbf{m}x + \mathbf{b}}$ y intercept

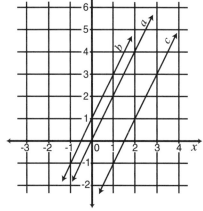

Line a $y = 2x$ slope? ____ y-intercept? ____
Line b $y = 2x + 1$ slope? ____ y-intercept? ____
Line c $y = 2x - 3$ slope? ____ y-intercept? ____

How do the slopes of lines a, b and c compare? How are the graphs of lines with the same slopes related to each other?

Nonvertical lines which have the same slope are parallel.

1. Predict which equations will have parallel lines as graphs. _____

   A $y = \dfrac{1}{2}x + 3$ **B** $y = 4x - 5$ **C** $y = -3x$ **D** $y = \dfrac{1}{2}x + 2$

List all the linear equations in each set that are parallel lines.

2. _____

   A $y = \dfrac{1}{2}x + 3$

   B $y = -\dfrac{1}{2}x + 4$

   C $y = 3x + 3$

   D $y = \dfrac{1}{2}x - 1$

3. _____

   A $y = 4x - 5$

   B $y = 5x - 5$

   C $y = \dfrac{1}{5}x$

   D $y = 5x$

4. _____

   A $y = -x + 1$

   B $y = x - 1$

   C $y = x + 1$

   D $y = \dfrac{1}{2}x - 1$

5. What is the slope of a line parallel to $y = \dfrac{2}{3}x - 7$? _____

6. Which equation has a graph parallel to $y = 2x + 3$ and through the point (4,6)? _____
   A $y = 2x - 2$ **B** $y = -2x + 3$ **C** $y = 2x + 4$ **D** $y = x + 3$

7. Which equation will be parallel to $y = \dfrac{1}{3}x + 3$ and pass through (9,2)? _____
   A $y = \dfrac{1}{3}x + 9$ **B** $y = 3x + 3$ **C** $y = -3x - 6$ **D** $y = \dfrac{1}{3}x - 1$

Slopes of Perpendicular Lines

Perpendicular lines lie in the same plane and intersect at right angles. The corner of a sheet of paper is a real-world example of perpendicular lines.

Line g: $y = 2x + 2$ Line h: $y = -\dfrac{1}{2}x - 2$

The equations of lines g and h are written in the slope-intercept form of $y = \mathbf{m}x + \mathbf{b}$, where **m** is the slope and **b** is the y-intercept. Tables of coordinate points for these two lines and a graph of the lines are shown.

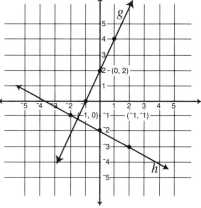

line g

x	y
-1	0
0	2
1	4

line h

x	y
-2	-1
0	-2
2	-3

How are the two lines related?

How are the products of the slopes of perpendicular lines related?

> The product of the slopes of two perpendicular lines equals -1.

1. Match the sets of perpendicular lines.

_____ $y = 6x + 3$ **A** $y = -\dfrac{3}{2}x + 5$

_____ $y = \dfrac{2}{3}x - 4$ **B** $y = -\dfrac{2}{3}x - 3$

_____ $y = -3x + 1$ **C** $y = -\dfrac{1}{6}x + \dfrac{1}{3}$

_____ $y = \dfrac{3}{2}x - 6$ **D** $y = \dfrac{1}{3}x - 2$

2. Find the pair of perpendicular lines. Tell how you know.

A $y = \dfrac{4}{5}x + 3$ **B** $y = \dfrac{5}{4}x - 2$

C $y = 5x - 1$ **D** $y = -\dfrac{5}{4}x + 1$

Lines _____ and _____ are perpendicular because _____.

3.

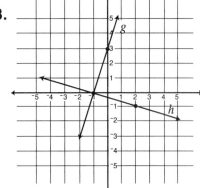

What is the slope of line g? _____

What is the slope of line h? _____

Are the lines perpendicular? _____

How do you know? _____

_____

4.

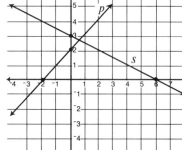

What is the slope of line p? _____

What is the slope of line s? _____

Are the lines perpendicular? _____

How do you know? _____

_____

Writing Equations from Data

Tables and graphs represent data. Linear equations also represent data that increases or decreases at a constant rate.

Example: Write an equation for the data.

A parachutist jumps from a plane. The chart below shows the rate in feet per second at which he is falling before he pulls the rip cord. Write an equation to show the relationship.

time (t) in seconds	1	2	3	4
rate (r) in feet per second	32	64	96	128

Look for the pattern between t and r.

$$r = t \cdot 32 \text{ or } r = 32t$$

Another way to find this relation is to find the slope of any two ordered pairs.

$$\text{slope} = \frac{y_2 - y_1}{x_2 - x_1} = \frac{64 - 32}{2 - 1} = \frac{32}{1} = 32$$

Use the slope and one of the coordinates to write an equation.

$$32 = \frac{y - 32}{x - 1}$$

$$32(x - 1) = y - 32$$
$$32x - 32 = y - 32$$
$$32x = y$$

Write an equation for the sets of data. Graph the data.

1. Time conversions

Weeks (x)	1	2
Days (y)	7	14

Slope = _____ Equation: _____

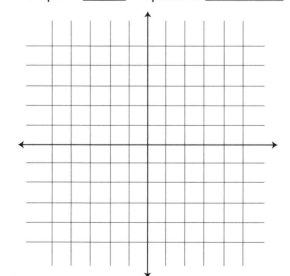

2. DVD's purchased

Number (x)	1	2
Total price (y)	$20	$40

Slope = _____ Equation: _____

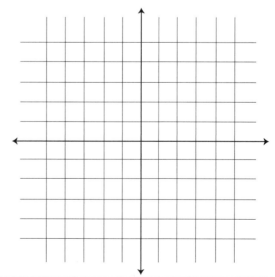

Solving System of Linear Equations by Graphing

Problems involving more than one unknown may be solved by using a separate letter to represent each of the unknowns. If a problem has two unknowns, it may be solved by writing two letters in two equations.

Example: Find two numbers whose sum is 12 and whose difference is 6.

Let x = the larger number

y = the smaller number

$x + y = 12$ $x - y = 6$

x	y
2	10
4	8
6	6

x	y
0	–6
2	–4
4	–2

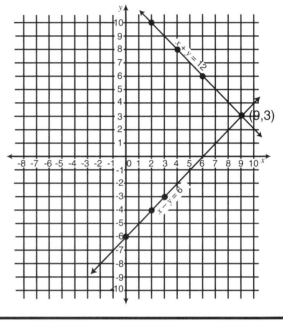

Graph both lines.
Look for the point of intersection.

The point of intersection is (9,3)

Check: $x + y = 12$ $x - y = 6$

          $9 + 3 = 12$ $9 - 3 = 6$

            yes   yes

Graph each of the following pairs of equations on the same grid. Check the solution in both equations.

1. $x + y = 5$ $x - y = 1$

x	y

x	y

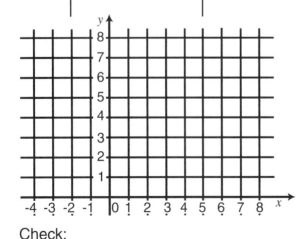

Check:

2. $x - y = 2$ $x + y = 4$

x	y

x	y

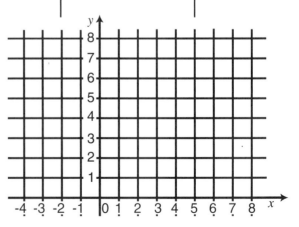

Check:

Solving Systems of Linear Equations by Addition-Subtraction

A pair of linear equations can be solved if they can be combined to form a third equation which has only one unknown.

Example: Find two numbers whose sum is 14 and whose difference is 2.

Let x = larger number

y = smaller number

$$x + y = 14$$

$$+ \ x - y = 2$$

Since there are the same number of y's, add the two equations together to eliminate y.

$$2x = 16$$

$$x = 8$$

If $8 + y = 14$,

$$y = 6$$

Check by substituting values of x and y into both equations:

$$x + y = 14 \qquad x - y = 2$$

$$8 + 6 = 14 \qquad 8 - 6 = 2$$

$$\text{yes} \qquad\qquad \text{yes}$$

Solve the systems of linear equations by addition or subtraction. Check.

1. $x + y = 1$　　Check:

　　$x - y = -5$

2. $x + y = 2$　　Check:

　　$x - y = 4$

3. $x + y = 8$　　Check:

　　$-x + y = 2$

4. $x + 5y = 12$　　Check:

　　$x + 4y = 10$

Hint: Subtract the second equation from the first.

5. $y = 3x - 5$　　Check:

　　$y = 2x$

6. $y = 2x + 4$　　Check:

　　$y = 3x + 6$

Graphing Quadratic Equations

The degree of an equation is the same as that of its highest-degree term.
    First degree equations: $x + y = 4$ and $4x + 8 = 0$.
    Second degree equations: $x^2 = 36$ and $y^2 = 25$.

A second-degree equation is called a **quadratic** equation.
Quadratic equations may be solved by graphing.

Graph of $y = x^2$

Example: Solve the equation $y = x^2$.

Make a table for x and y.
Graph the points.

x	y
3	9
2	4
1	1
0	0
−1	1
−2	
−3	

Complete the table. Plot the last 2 points on the graph.

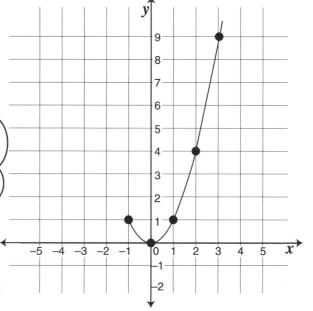

This curved graph is called a **parabola.**

Graph each equation.

1. $y = -x^2$ x | y

2. $y = 2x^2$ x | y

3. $y = x^2 + 1$ x | y

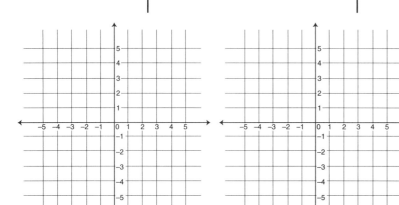

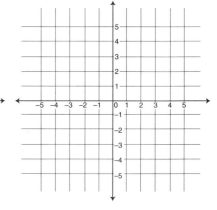